Lessons Learned Selling HVAC Service
Copyright © 2017 by Ray Wohlfarth

If you have any questions or comments, please send them to
Ray Wohlfarth
www.FireIceHeat.com
834 Kerry Hill Drive Pittsburgh, PA 15234
Tel 412-343-4110
ray@fireiceheat.com

ISBN:
ISBN-13: 978-1545544907

DEDICATION

This book is dedicated to Bob Priganc, Herm Dieckmann, and Ken and Rosemary Launikonis, my favorite sales trainers, and Dan Whalen, a wise man who took a chance on me. Whether you are starting in the sales side of the business or a grizzled old veteran, I hope this helps lead you to success.

Table of Contents

ACKNOWLEDGMENTS

I would like to acknowledge the following people who
supported me while I wrote this book:
Sheila, Jon, Ryan, Abby, and Conor Wohlfarth
Dan Holohan and Erin Holohan Haskell
My writing group friends for their support and suggestions:
Tim Reilly and Keith Bastianini
My Creative Writing teacher, Dan Shapiro
RIP my friend D. Brian Baker, who would challenge me and
made me a better writer.

About You

How does someone get into HVAC sales? A common way is when a service technician is asked to start doing sales. Another avenue to become an HVAC sales person is because the person could not get a real job. Our industry is not considered glamorous to outsiders and few flock to it but it has been terrific for many sales professionals. Some are thrust into sales when they say *Enough is enough* and make plans to start their own business. Perhaps they worked for a knucklehead or just felt they could do a better job than the current employer. In any event, every entrepreneur knows sales are the key to growing a business.

My HVAC career started as an apprentice with the local Steamfitters union under the supervision of journeyman steamfitters. I was drawn into the industry when my dad said: *"People are always going to need heating and cooling."* And he was correct. In my last year of my apprenticeship, I was involved in a serious vehicle accident on the way home from a job site and unable to work in the field. A local contractor took a chance on me and offered me a sales position.

Selling service is more complicated than selling an item the customer can see, smell, or taste. I used to be envious of sales professionals who sold vehicles which I incorrectly believed sold themselves after the test drive. When selling an intangible such as service or maintenance, it takes some creativity. While we cannot sell the new car smell or the way the equipment can corner, we can sell on things such as safety, efficiency, carbon footprint, comfort, fewer complaints, or reduced breakdowns. These can be equally as persuasive in the customer's mind.

The following are some lessons I learned while working in the field for three decades. Some of my ideas may resonate with you and others may not. I want to help you avoid some of the many mistakes I made. The book will focus on how to increase sales.

Treat your sales career as a business

I learned this early in my sales career, treat your sales career like it is a business. Think of it like You, Inc. What does that mean? You, as a business, are selling your services to your employer. This arrangement may be short-term or long-term. Try to learn all you can from the company where you work and accept any training offered as you are looking for the long term health and wealth of You, Inc. If the employment does not last as long as you expected, you are now better for your time there and provide more value to the next employer. I worked for a contractor and loved it. The manager treated me great and with respect and I worked hard for the company. One Monday morning, I walked in and met my new manager and he explained the previous boss was gone. The new boss had impressive educational credentials but not one bit of experience with sales. He thought all sales people were shysters and scam artists and he treated me that way. I took Ray, Inc. to another firm and learned all I could there. When I started my own company, I had experience from the previous firms and it helped significantly.

Have fun

When I gave the two-week notice to resign and start my business, the manager shook my hand and said how excited he was for me and wished he could start his own business. The manager was under constant pressure to cut costs and increase profits. I suggested he could start his own company and he said: *"I only have 15 more years until I retire."* I thought if a prisoner were sentenced to 15 years with the stress he was under, they would say it was cruel and unusual punishment. In spite of his hard work, my old boss lost his job a few years after I left. This drove home a powerful lesson, always do what you love. A wise man once told me if you do what you love, you will never work a day in your life. It has been several decades in the industry and I still love getting up each morning and doing my job. I love the fact I can learn something new each day and help people. Try to find fun in your job.

Honesty and Ethics

As you begin your sales career, one of the things I implore you to think about is always to be honest with the customer. It is such a close-knit industry and reputations earned are difficult to change. You never know where the industry will take you so always use the truth. It is ok to tell the client you do not know the answer to his or her question. I would follow it up with *"but I will find the answer."* You will earn the respect of the customer.

I was taught by a wise journeyman always to work as if you were being filmed from a hidden camera. In this time of security and cell phone cameras everywhere, it is not far from the truth. He told me this to help me remember always to do the right thing. I offer this advice to you as well.

Be thankful

I try to find five things to be thankful of each day. Some days are easier than others to see those items. I do not mean to sound metaphysical but believe you attract what you think about in life. When you are on this emotional roller coaster called sales, it can quickly lower your confidence when you lose a big sale. The fastest way to pull yourself out of the funk is to concentrate on positive things in life.

Why a sales career is great

Sales, as a career, has many positives. Your schedule is mostly your own and the boss usually looks the other way as long as the sales are coming. I was able to watch my children's sports events during the day. Do you want a raise? If you make one more cold call per day, your sales and income will increase.

Who are you?

In your career, you will have to make choices about whether or not to do the right thing. It is not always an easy choice. One of the things that kept me grounded was one of the favorite poems, The Man in The Mirror. While written in 1934 for a man, it could just as easily apply to anyone.

The Man in the Mirror

Peter Dale Wimbrow Sr.

When you get what you want in your struggle for self
And the world makes you king for a day
Just go to the mirror and look at yourself
And see what that man has to say.

For it isn't your father, or mother, or wife
Whose judgment upon you must pass
The fellow whose verdict counts most in your life
Is the one staring back from the glass.

He's the fellow to please – never mind all the rest
For he's with you, clear to the end
And you've passed your most difficult, dangerous test
If the man in the glass is your friend.

You may fool the whole world down the pathway of years
And get pats on the back as you pass
But your final reward will be heartache and tears
If you've cheated the man in the glass.

How to stay positive

Sales can sometimes erode your confidence and each sales person has to discover a way to stay optimistic. I have had my share of sales calls I thought I would close only to find out they chose another firm. Try not to take it personally. I realize it is sometimes more challenging than it seems. I like to take a walk in nature to help me refocus and purge the negative thoughts. Some use exercise to get rid of the feelings. A key I use to deal with the ups and downs of a sales career is to be steady. Try not to be too high when it goes your way and try not to get too low when things do not work out for you. I also like to look at the odds to help me get past it. For example, an industry rule of thumb suggests for every ten cold calls we make; we will have 2-3 appointments. We should be able to sell one person from the ten sales calls. I like to look at it and say, thank you to the individual who turned me down as it takes me one step closer to a sale. Stay positive.

Feel-good drawer

When life throws me a curve and seems unfair, I open a special desk drawer I call the feel good drawer. No, it does not have a bottle of whiskey. It has accomplishments I have received in my life. It contains hand-written thank you notes I received in my career. There is also a copy of my first magazine article and my first book cover. There is a picture of my children. The feel-good drawer is a conglomeration of your successes. It is there to make you feel better. Whenever you get an award or a pat on the back, save them and put them in a safe place. You can read them when you are feeling down.

Chasing **DOLLARS**

Several years ago I was asked to speak to the graduating class of a large trade school in my area. The person who invited me to speak challenged me to come up with something which was easy to remember and they could keep it throughout their career. I know most people spend their career chasing money and used it as a benchmark for their level of success. I used DOLLARS as an acronym for keys to use through the career.

Differentiate yourself

You want to separate yourself from the others in this trade by being slightly different. It does not mean it should be different in a weird or bad way. It means to be different in a positive way. Things like saying Thank You and Please will separate you from the rude, self-absorbed people.

Own you're your career Don't just rent it

Many people make the mistake of doing as little as possible and renting their career for eight hours or less per day. If you own your career, you will use some of your spare time to make yourself better by reading, watching training videos, or attending classes. In addition to mentally training, you should consider physically training. Sales, although not as physically demanding as working with the tools, will be mentally draining. Try to do some cardio to rid your body of the stress.

Learn all you can

I urge you to learn all you can about the trade. This education could include sales training, technical training, and learn about your customer's industry. For example, if you work with apartment or rental agents, try to learn some of their terms so you can talk with them in the industry language.

Lean on someone

If you can find yourself a mentor, it will make your path to success quicker and less bumpy. The counselor will help you avoid mistakes. Many would love to be asked for their advice. A friend of mine who was a one man company asked a retired successful business person to be on his Board of Directors. This Board Meeting amounted to meeting for coffee and discussing what he was doing and listening to the mentor's suggestions. It helped my friend grow his company over time.

Allow yourself to make mistakes

Many are devastated when they make a mistake and it is ok. It happens to everyone. The only people who do not make a mistake are the ones who do nothing. Willie Stargell, famous baseball player for the Pittsburgh Pirates, had 48 home runs in 1971 to lead the league and his team to the world series. That same year, he had 154 strikeouts which meant he failed three times more than he succeeded. Willie Stargell is recalled for his hitting prowess and not his strikeouts. You will be remembered for your successes and not your failures.

Read all you can

If you want to separate yourself from the pack, try reading a little bit each day. It doesn't have to be much because most people never read another book when they finish school. I urge you to try reading at least 15-30 minutes a day. Try reading about companies in different industries and how they go to market. You may get some ideas for your company. I like to read in the morning before the phone rings while some people prefer reading at night. A friend shuts off the phones at lunch and reads a half hour each day. What should you read? I like sales books, technical information, or trade magazines. Once you get into the habit, you will enjoy it. Another method of gaining information is to listen to audio recordings in your vehicle or while exercising.

Specialize

Try to find a niche in the industry you like and become the expert. It will allow you to have higher margins and better success. We specialize in boilers and there are only a few companies in my area who know about them. It is a great feeling to walk in the door and fix something two or three companies were unable to do. When we do traditional heating and cooling, we are one of several hundred HVAC companies in the area and it is hard to get higher margins and separate ourselves from the others. Remember, the lead horse is the only one that gets a change of scenery.

Enthusiasm

Just how enthusiastic should we be? I was teaching a sales class and one of the students had this annoying way of being overly enthusiastic. It is great being enthusiastic, but this guy was over the top. When someone would ask him how he was doing, he would boom out, *"Fantastic and why shouldn't I be?"* He would shake your hand aggressively and invade your space. It was almost intimidating. One day after the class, we all stopped for an adult beverage and he and I sat next to each other. He looked at me and said *"I may be losing my job. My boss gave me a 60-day notice I will be gone unless my sales increase. I don't get it. I'm enthusiastic and smile but people still do not buy. I have a baby on the way and need this job."* I felt sorry for him and asked if he would like my advice and he nodded. I told him, while it is great to be enthusiastic, it can sometimes be intimidating. I suggested he try being more like a consultant or doctor rather than a sales person. I used the analogy of going to a doctor appointment with a pain in the side. Could you imagine seeing your doctor jump up and aggressively grab your hand and shake it? What if the doctor said *"Hey, I am having a huge Labor Day sale on kidneys. I can get you no interest financing for five years. What do you think? And by the way, I can give you an additional 10% off if you sign today. What do you think?"* I would run out of the exam room as would you. Instead, the doctor takes x-rays and tests and then talks with you about what the tests showed. I explained this is how a professional sales person should act. The young salesman agreed to try it. The next week at the class he showed up with a huge smile on his face and said *"That really works. I got two sales."* Although he was still positive, he was more subdued. Even though sales books will tell you to be enthusiastic, customers are sometimes intimidated by an overly enthusiastic sales person.

Finger over lips

As a repairman, I like to repair things. I had to learn to bite my tongue when I made the move into sales. Some things are best left unsaid. When a client is talking, I have the urge to open my big mouth and solve their problem. I will fold my index finger and press my knuckle against my lips to remind me to be quiet. One of the things I learned after being married is sometimes people do not want a solution, they only want to vent. If you try offering a solution to someone venting, it does not end well. I will jokingly ask, *"Would you like a solution or are you just venting?"*

Do the right thing

Sometimes the job does not go as expected. You are in a difficult quandary as doing it the right way will sometimes cost you or your organization money. What do you do? If it is a small amount, I may just absorb it and let the client know how I went over and above for them, like an IOU. If I find it is a significant amount, I like to meet with the client and discuss the findings. I will explain what we plan to do to resolve the issue. Most customers understand there are circumstances beyond your control.

We had a building owner call us to perform a clean and check on his furnace. I had a weird feeling about the owner as he pointed to the basement door and told me the furnace was downstairs. Most people will accompany you downstairs to help you navigate the basement, but he stayed upstairs. *Always listen to those feelings, by the way.* When I walked to the furnace, I saw it was not working. I walked upstairs and asked the man about the furnace. He got angry and said it was working before I got there and I must have done something. I was shocked and informed him I simply saw the unit was not running even though there was a call for heat. In an attempt to diffuse the situation, I offered to diagnose the cause of the issue. The problem was a defective thermostat. After informing the customer of my diagnosis, he started yelling and said I was blackmailing him and should replace the thermostat for free as it worked before I arrived. I offered him a discount and the customer was adamant about getting it free. I felt the anger building and walked off the job site. The customer refused to pay my invoice for the service call. I found out later this man had a reputation for doing the same thing to other service companies. Sometimes doing the right thing is not enough.

It's in the details

I urge you to add a disclaimer which absolves you of items found which are beyond the scope of the proposal. For example, you are replacing a furnace and plan on reusing the existing thermostat wire. Once you get on the job site, you find a broken thermostat wire inside the wall and have to install new wire. That costs would be hard to absorb. I have a detailed list of the items we will reuse or replace on my proposal to eliminate or limit the gray areas.

Making you better

Famous NFL football coach Chuck Noll used to say *"It's all about blocking and tackling."* In other words, you have to be proficient in the basics of your craft to become successful. Each year, professional athletes attend a preseason training camp where they work on the fundamentals of their sport. Most of these athletes have been playing since they were a small child and yet do this training yearly. It is a good idea to attend training and review the basics at least once a year. There are many local and national seminars to attend which can help you hone your skills.

Sales Training

You would never allow someone with no experience to work on your vehicle and you should use that same philosophy with sales. Sales, like any other career, requires training. There are some excellent sales training assets where you can learn about the sales. These may be anything from sales training books to classes hosted by a local supply house or industry organization. If you like videos for learning, YouTube has hundreds of "Sales Training" videos. I figured if it took me four years to become a journeyman, I should invest that much time in sales training. I read books, listened to sales recordings, and attended sales training classes. Some of my favorites were Dale Carnegie, Sandler Sales Training, and Brian Tracy.

One of the frustrating things to me is some HVAC business owners expect their technicians to sell without one bit of training. I think it is an insult to the trade and an insult to the employee. I believe technicians should look around for legitimate opportunities while on a service call.

Look at companies you like to use

Look at successful companies which are not in the same industry. I ask myself, What do they do to earn my business? There is a small restaurant I enjoy visiting, Someone Else's Bar. The owner, Else, is there each night and greets everyone with a smile and knows your name, even if you have only been there once or twice. The food is excellent and although it is slightly more expensive, I will go there rather than a large national chain. Why do I like going there? She appreciates my business and thanks me every time for choosing her restaurant. I do the same with my customers.

How Much Training

Famous writer and consultant Malcolm Gladwell has the 10,000 Hour Rule which states to be world-class in any field requires 10,000 hours of deliberate practice. It seemed a bit extreme to me as I had a family and a life. I subscribe to Ray Wohlfarth's Horse Racing Philosophy of Training which states a win by a nose is still a win. All you have to be is a small percentage better than your competition to win. Did you know 33% of adults never read another book after school? I tried fitting my training around my schedule. I am a morning person and read about an hour a day before work. I also listened to self-help recordings while driving to and from job sites. If you are going to specialize in an area, try getting as much training as you can afford on your subject. You may be able to find inexpensive training for your niche through the local supply house or manufacturer's representative. Industry organizations such as RSES, ASHRAE, or OESP may have local or regional training at little or no cost. Many of the manufacturers also offer training at their facility.

Visit ASHRAE AHR Expo

Since you are in the industry, I urge you to visit the ASHRAE AHR Expo which occurs each January. It is in different cities every year and filled with vendors and manufacturers of HVAC equipment. You can see what is new in the industry. If you have never been to the expo, it will amaze you at the size and breadth of our industry. Also, the expo will have hundreds of training seminars on just about any subject you can imagine.

Community College

Many of the local community colleges have noncredit courses to make you a better sales or business person. I have taken classes on writing, book keeping, and software. Besides, you could meet possible new customers. I highly recommend the writing classes as it will help you write compelling letters, quotes, and sales flyers.

Understand Finances

When you are selling commercial HVAC equipment, you will most likely be speaking with the finance people. Consider enrolling in classes on finance so you can explain how your HVAC solution can help the customer. If you can talk in financial terms, it will increase your sales. Many of these classes are available on line or at the local community college. Building Owners and Managers Association or BOMA has classes to become a property manager. If this is your target market, you could attend the classes and meet property managers.

On Line Training

Many equipment manufacturers have on line videos where you can learn about their product. There are other places such as You Tube where you can watch videos and learn about different products. There are also free college classes available on lIne.

If you don't like something, change it. If you can't change it, change your attitude. Maya Angelo

Role Play

When I first started working for the local branch of an international company, the boss believed in role playing. He would take the role of the demanding customer and I was the eager sales person. While I balked at this type of training, it did help to prepare me for some of the people I met while doing sales calls. If you can have someone role play with you, it will help you once you get in front of a difficult prospect. I used to have my spouse help because she could be a tough sale.

Find a Mentor

The Small Business Administration or SBA has some interesting videos you can use to learn about how to operate a business. Also, the local SBA has a program where they will connect you with a mentor to help consult with you and it's free.

Another way to find a seasoned pro is to talk with some of the wholesale houses. They may know some retired people dying to get out of the house and share their knowledge with you.

Your local college or university business department is another source of free advice. The students in the MBA program here offer free consultations and it helps them get some real world experience.

Express yourself professionally

Some techs do not interact socially very well and as a result, have a difficult time speaking with the customer. I urge you to practice speaking into the mirror as a way of hearing the way you speak. Another way to hear how you sound is to record yourself. It sounds hokey, but it actually works. To be successful, you have to communicate well. The customer is going to look to you for your expertise and recommendations.

Writing is another area where you have to communicate well. Before you send a letter or email, try reading it aloud. You will hear how the letter sounds and whether changes should be made before sending.

Spell Check

One of the great mysteries in our industry is the most gifted service technicians seem to have the worst spelling/grammar skills. If your service report has spelling and grammatical errors, the customer will judge you and your ability. I knew a contractor who would ask job applicants to complete a typical service report to see how they wrote. I used to furnish dictionaries to my technicians but it did not work very well as most were placed on a shelf in the rear of the truck and never opened. We switched over to electronic tablets and the service software has a spell checker. I was in heaven. In addition, I added a list of the common problems and solutions in a drop-down list so the tech could finish the service report quickly. It worked out well for me and the techs.

Don't criticize the installer or designer

One of the mistakes some techs make is to criticize the original designer or installer. Some technicians feel like they should redesign every system. It could anger the customer and leave you liable for a lawsuit. I tell my employees to see if they can make the existing system operate properly.

During an unusual cold snap, we visited a school district office which was cold. When my technician arrived, he looked at the fin tube radiation and declared it to be too small and told the owner there was nothing further he could do and left, proud of himself. The first thing the owner did was to call the architect for the district who in turn called the mechanical engineer who in turn called me. It was not a pleasant call. The engineer screamed that he was going to sue me for defamation. I contacted my technician and he agreed with his original diagnosis. When I explained the system worked great for ten years without an issue, he stood firm with his diagnosis. I drove to the job site and checked all the basics and found the pump was too small. The maintenance department for the school had recently replaced the pump with an undersized one because it was less expensive than the properly sized pump. We replaced the pump with the properly sized one and the system ran great. I apologized to the building owner and the engineer and avoided a lawsuit.

Cell phones on a job site

Every tech will have a cell phone and they can be a great asset. Be careful if it appears like you are on the phone during the service call. I had customers call and complain the tech was on the phone the entire time and wanted the bill reduced.

Give them to clients

I like to purchase extra company attire and give them to customers. It may be anything from hats to tees to even jackets. It is an inexpensive marketing tool especially if the person is an influential person in the market. It is like getting a famous athlete to wear your shoe.

Any type of exposure?

This is a business philosophy I have heard in my career but am not sure about the validity of it. We interviewed a technician and decided not to hire him. There was something about him which did not feel right. A few months later, I saw the man I interviewed arrested for robbing a convenience store. The news showed the van of the company he was working for and the logo on his uniform. I was glad I did not hire the man.

Be careful wearing the company attire to somewhere you may not want the attention. An employee of a competitor who was wearing his company logo on his shirt got drunk and obnoxious while attending a trade show event. His dumb behavior did unknown damage to the company.

While I was an apprentice, one of the journeymen technicians drove his company vehicle to the local strip club and parked in front. The company logo was there for anyone to see. The next morning, the boss received a call from a customer complaining about seeing one of the trucks at the club and asked what kind of company we had. It was a bit of an overreaction but how many other customers or potential companies rode by that night and thought the same thing. A friend of mine who serviced beer coolers at these types of establishments told me he had the same problem and either parked the vehicle in the next lot and walked to the bar or hid his van behind the building.

Differentiate yourself

I read a study which claimed most horse races were won by a small margin, Hence the term, Winning by a nose. Consider your career like a horse race. If you win by a nose, you still win. All you need to be is slightly better than your competition. Try to differentiate your company positively.

It is small things which make the most difference. Little things like saying please and thank you mean a lot to existing and potential clients.

Don't take them for granted.

Let your customers know how much you appreciate their business. In addition to having it printed on every invoice, I thank the customer face to face or on the telephone for their business. You will be amazed how effective it is. No one says thanks to people for their business.

What is your job?

Many sales people think it is their job to judge the client for their lifestyle, political choice, sports team, or their decision-making ability. Your job is to provide a solution from your company to help them with their heating and cooling. If you want to make money, stick with heating and cooling advice. If you want to have a sales call go south quickly, try talking politics.

Are you the backup company?

We worked for a real estate management firm and was their backup HVAC service company. The primary company, which consisted of one person, would handle all the calls during the day because his labor rate was about $10.00 per hour less than ours. In addition, he would sell them parts at his cost, never a good idea. At night, the man would turn off his phone and they would call us. The customer was very demanding and expected a prompt response regardless of the hour. The service calls were usually in dangerous parts of town and not someplace you would like to go alone. On one midnight call, the house had several guys with guns seated on the couch and it was scary. One of the guys led me to the basement and my heart raced the entire time. After cleaning the flame sensor and getting the heat running, I bolted out of there. I met with the owner of the real estate firm and explained what happened and he laughed and said, *"They are not a bad group."* I asked if we could get some of the better work such as replacements or at least daylight work. He told me we could if we cut our rates by at least $10 per hour. I respectfully told the man we could no longer service his buildings and walked out. My life or the life of my techs were not worth the risk.

Our greatest weakness lies in giving up. The most certain way to succeed is always to try just one more time. Thomas Edison

Know your products

When selling any product, you should know your products and services better than your clients. Read the technical and sales information on the products. Many customers use the internet to research products and services.

Plan your work, work your plan

To be successful in sales, you should develop a marketing or sales plan and stick to it. Sometimes, you may be tempted to change your plans, especially after a few bad days in a row. You need to believe in your plan and keep working it. Do not allow emotion to guide your decisions. I had two large school boiler projects we were assured to get. These sales would have made my entire year. Within a week, we lost both projects. I was devastated. While assessing what happened, I made a vow to myself to never sell to schools again. It was an emotional response and after calming down, I went back to selling to schools, one of my niche markets. The loss motivated me further and I worked harder at getting sales and grew my market share. The only way it happened was because I worked my plan.

Motivation will almost always beat mere talent. Norman Ralph Augustine

How do you know if the plan is working?

This is a common question when evaluating a sales plan. The most obvious way to test if a sales plan is working is by the sales you get. When selling an item with a long sales cycle, it may take a year or two to develop and may be harder to track. The way you track these type of sales is to see if you are progressing toward a sale. For example, let us look at the steps involved with a long term sale.

1. Meet with client and explain that the unit should be replaced.
2. Discussion of the budget prices.
3. Approval of the funding.
4. Sales proposal and the client authorization.
5. Order the equipment.
6. Installation and startup

Let us say you met with the owner and discussed the problem and gave them a budget price last month. The next step is to verify the funding is approved. If it is, you are making steps to the finish. One of the nice things about these types of sales is once you get going, you may get sales from efforts you did last year, almost like a bonus. I would suggest you go somewhere outside the office and evaluate how the sales plan is working.

Life is 10% what happens and 90% how you react to it.
Charles R. Swindoll

Understand systems

The successful companies in our industry understand HVAC systems and are not just parts changers. While you as a sales person are not expected to be a system expert, your knowledge of them could help you with your sales. If you do not know systems, be sure to talk with the technicians and bosses to help you understand them. You could enroll in the local community college and take classes on HVAC. You will also meet some potential customers in the class.

Do not diagnose while in the truck

I was guilty of diagnosing HVAC problems many times as a technician while on the ride to the job site. Once I made up would make up my mind, I would get stuck if it was something other than what I thought. This habit continued once I started in sales. I recommend you walk in and look around the system without any preconceived ideas and see what the problem is. I dropped my vehicle at a repair shop and told the mechanic what I thought was the problem. He looked at me and said, *"Don't tell me what you think is wrong. I want to diagnose it myself."* I was taken aback at his brisk nature but he was right and my solution was not the correct one.

In the business world, the rearview mirror is always clearer than the windshield. Warren Buffett

Your time is worth money

Many techs or sales people make the mistake of doing unbillable or under billable work during the day. First of all, I admit I am a cheapskate. I had to replace a printer for my office and spent several hours looking for one. I drove all across town to save a few dollars. Remember two things: You get paid for your time and there is a finite amount of time in a day. If you charge $100 per hour for your expertise and labor, be sure you are doing $100 per hour work. Sometimes we get caught up doing work below our pay grade such as cleaning the office, shopping for a printer, or changing the oil on a vehicle. Always ask yourself *"Is this $100 per hour work?"* If not, hire someone to do it and find more $100 per hour work. Let me use the example of my printer. I spent several hours on line and traveled to the store across town and back to get the printer. All of this effort to save $40.00. But did I actually save $40.00? I lost four hours of billable labor @ $100 per hour for a total of $400.00. It cost me ten times the savings in lost earnings. This translates to sales as well. You should spend your time making sales calls during the day. Don't trip over dollars to save pennies.

Not my job

When I first started in the trade, I worked for a pneumatic control company and would service and install the controls in commercial buildings. I was sent to a job site and find my controls to be working properly and the issue would be in either the electrical or mechanical system. The customer would have to contact another vendor to come in and provide service on the electrical or mechanical side. It led to finger pointing and angry customers. One of the things I did was to attend electrical classes at the local community college. I was able to repair most of the issues because as you know, most problems are on the electrical side. You need to be able to provide a one stop shop for your customers.

How to keep going

I was talking to an insulation contractor at a trade show a little while ago and said I had not seen him in a while. He smiled and told me life had dealt him a couple of blows and it took him a few years to dig himself out. A contractor he worked for suddenly went out of business and stuck him for $140,000. I know many people would have quit but he kept going. Some days are better than others in this industry and we all have a bad day or bad week once in a while. The key to overcoming a bad day is how you respond. You have two choices, quit or fight through. It is sometimes tempting to chuck it all and walk away. I guarantee you if that happens, you will regret it. My friend Dan calls those Life's Lessons and hopefully, you will learn from them. Prayers also help when you are knocked down.

Need to make a decision? Try fishing

I read an article where the author suggested fishing when you have an important decision to make. The author believed the idea of being away from the office in nature would help you to ponder your problem in a better environment. I have tried it many times and found it to be very beneficial. My spouse did not agree with the remedy when we had toddlers at home.

Visit Industry websites

I like to visit industry specific sites such as HeatingHelp.com to get ideas and helpful hints. It also allows you to get real life tips.

CYA

The acronym CYA means, Cover your A**. You have to protect yourself and your company in business. I had a project where the design engineer specified an on off burner for the project. We sent an email to the engineer informing him the burner was an On-Off burner. His return email confirmed that was the one he wanted. During the construction phase of the project, the engineer went to work for another firm. The owner's representative threatened us with a lawsuit because he did not want an On Off burner. We were able to show the email and it saved us both our reputation and lots of money. I urge you to protect yourself with emails to document your conversations.

Get the price in writing

I called a vendor and asked for a price on a piece of equipment for a quote I was working on for a customer. The salesperson told me the price and I used it in my quotation. We were awarded the contract and when I ordered the equipment, the price was $2,000 higher than what I was quoted. I argued with the sales person and he denied giving me the lower price and we had to eat the $2,000. Two things I learned that day; I never used that firm again and I always get the price in writing.

Some employers lie

I worked for a small mechanical contractor in my area when I started in sales. I will be forever grateful to him for taking a chance on me and showing me how to become a sales professional. I was approached by a large international company after I was there for a few years and asked to interview with them. During the interview, I was promised a substantial raise if I went to work for them. It turned out to be a lie. I took a pay cut because of the promised bonuses and commissions. They never came and when I asked the boss about it, he shrugged his shoulders and said it was out of his control. Get all promises in writing.

Bring over all your clients

I interviewed for a sales position with another company and felt a weird feeling in the pit of my stomach. The two people interviewing me asked questions about how my current company did business and wanted me to share confidential information about my existing employer. I excused myself and told them I was not interested. I could never do that to a company which hired me and provided for my family.

My experience with Dale Carnegie

When I started in sales, my boss sent me to a Dale Carnegie Sales Course. It was very informative and I learned about the sales process. After graduation, I was asked to return as a group leader and teach the new students. This is where I actually learned sales. I spent several years as a group leader and assistant instructor. It allowed me to do two things; it made me a better sales professional and I met many people who helped get me the HVAC service at their office. A way to make you a better sales professional is to teach a sales class at your company. If you teach the technicians about sales, it will also pay dividends.

I have not failed. I have found 10,000 ways that won't work.
Thomas Edison

Should you install equipment they purchase?

With people shopping on line, we are asked by customers to install equipment they purchased on line. I try to dissuade customers from doing it. Some wholesale houses will sell their equipment directly to the building owner and they will ask us to install it. I explain we cannot warrant the equipment the customer purchased. We installed a packaged rooftop unit the customer purchased at the supply house. The unit had several issues which required a couple of service calls. We invoiced the customer for the labor and he asked why it was not covered under warranty. I explained the labor was not under warranty because he bought the unit. This led to an angry client. They were not necessarily irritated at us but more the situation.

A different variation of this is to install customer furnished parts. Selling only labor is not a good idea. For one thing, you are assuming responsibility for the system using their parts. The other drawback to installing the customer furnished parts is a much lower profit margin for your firm. An industry rule of thumb suggests a labor to parts percentage of 80%. If you are selling labor at $100.00, you would typically sell $80.00 worth of parts. On a four-hour service call, you would generate sales of $400.00 in labor and $320.00 in parts for a total of $720.00. If you sold only labor, your sales would be $400.00 or 44% lower. Parts sales should be part of your sales offering.

Dealing with the difficult customer

I had a commercial client who would call the office and yell and scream if something happened to his unit. He would demand immediate service and say *"This thing is only a few years old."* The unit was ten years old. It is funny how customers lose track of time when they reference the last time you were there or the age of their unit. The key to dealing with the irate customer is to acknowledge their concern without taking it personally.

I would most likely say something like this:

"Mr. Customer, I can certainly understand your concern about what happened. If this happened to me, I would be upset also. Where do we go from here?"

This statement acknowledges the item which is upsetting the customer and it shows empathy for the customer. It further allows the owner to choose how to remedy the situation. In most cases, the customer only wants to be heard.

Fall on the sword

In ancient Japan, the Samurai warriors were known to fall on their sword if they were disgraced. Today, the phrase means to take responsibility for something you or your employee did. It is a powerful sales tool but limited in its use. For example, we completely missed a spring preventive maintenance call at a customer's building. The customer called me into his office. He explained what we did and looked at me. I took full responsibility and apologized for the mistake. I told him I would understand if he used another company. The customer felt satisfied and told me never to let it happen again. I assured him we would not. The drawback to this technique is you can only use it once. It loses its power if tried again. I use it only as a last resort.

The Pareto Principle

Italian economist, Vilfredo Pareto developed the Pareto Principle, a widely known economic principle which believes 80% of results come from 20% of the effort. He found that 80% of the land was owned by 20% of the people. In sports, 80% of all awards are given to 20% of the athletes. In our industry, you will most likely find that 80% of your business will come from 20% of the customers. 20% of the work force do 80% of the work. The key to growing your business is to find more of the 20% type customers and technicians while purging the high maintenance 80% clients and technicians.

Difference between needs and wants

When my children were young, they used to confuse needs and wants. For example, I would hear *"Dad, I need this article of clothing."* They disagreed when I explained the desire for the clothing was just a want and not a need. If you ever get a chance, do some research on Abraham Maslow. He was a famous psychologist best known for his theory of Hierarchy of Needs. It is fascinating as it shows the needs for any person from basic all the way to self-actualization. The basic needs for any human are food, water, rest, and heat, which is good for our business. The following page shows an example of Abraham Maslow's Hierarchy of Needs.

How does this help? If you know a basic physiological human need is warmth, your sales presentation could include how the new furnace could keep the family warm and comfortable. The next human need, safety is met by saying how the furnace is vented outside and the heat exchanger has a 20-year warranty.

"It ain't how hard you hit. It's how hard you can get hit and keep moving forward. It's about how much you can take and keep moving forward." Sylvester Stallone in Rocky Balboa

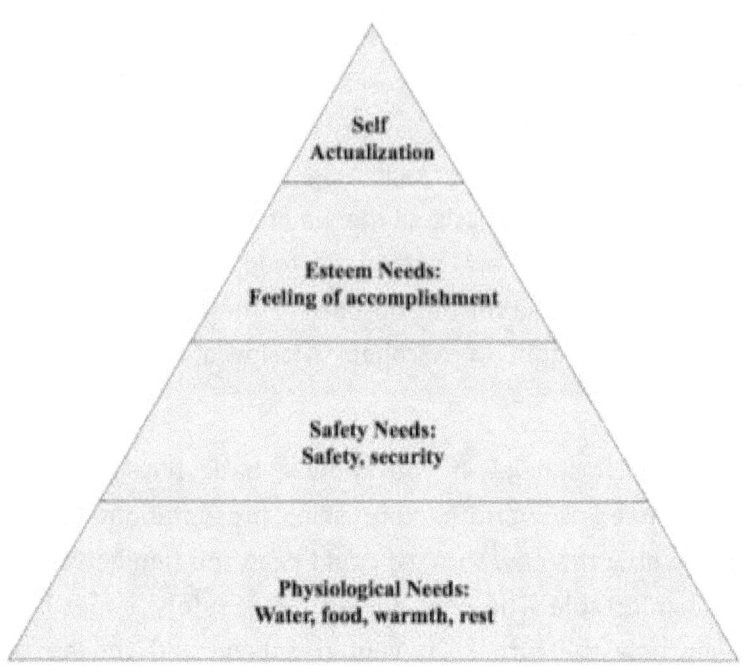

Holding the customer hostage

I met a manufacturer of custom made boilers and he had a unique system installed on his boiler control panel. The panel had a timer which would disable the boiler after 30 days of usage and the only way to start it again was to input a four digit code into the keypad. This was to assure the manufacturer he got paid in a timely fashion for his boiler. If the boiler were paid for, the manufacturer would tell the owner the code and the boiler would start. I felt uncomfortable holding the client hostage but the manufacturer said he tells the owner about it in the planning stage.

It's not Christmas

My friend Ken Launikonis used to say; *It's not Christmas* when dealing with prospects. He meant when you have an overly enthusiastic prospect, tell yourself "It's not Christmas." I once had a prospect call and said he heard about my firm and me and wanted to talk with us about doing his HVAC work. After inflating my ego enough, I met with the client and he told me he wanted us to do all his work. My Spidey Sense was screaming in my ear and I took a deep breath and remembered what Ken said. I asked the man who was currently doing his service now. He mentioned a competitor in my area with a pretty good reputation. He said something horrible about their company and said he would never have them on his properties again. I figured the account to be worth several thousand dollars year in revenue. I informed the client I would like to have my service manager with me while I looked over his equipment. He told me to hurry because he wanted us to do his work. As I left the meeting, I had this uncomfortable feeling and stopped by a nearby wholesale house to pick up a part. I asked my friend who worked there if he ever heard anything about the company and he looked me in the eye and said *"Stay away from that guy. Everyone who does work with him has to go to court to collect their money."* He told me of a small contractor who went out of business because the man refused to pay him for a unit he installed. I did some research and confirmed this customer would have been difficult to deal with and chose not to work with the person.

Recommend your competition?

I know the idea of urging customers to go to your competition most likely rankles you. Think of it this way, if you have a customer who is high maintenance or slow to pay, why not have them go to your competition? Let the competition spend their time and efforts trying to appease the client or collect their money while you look for quality customers. Think of it like pulling weeds in your garden so your plants can grow.

Avoiding a headache

We were looking for a service technician and set up interviews with potential hires. One person made me feel uncomfortable and I told him I would let him know in two days. I called his references and no one would talk with me about the man, which was odd. I received a call from one of the references and the caller said *"You don't know me but stay away from that man. He is bad news for your company."* I was going to tell the man I knew who it was because of my caller ID but allowed him to rant about the ex-employee. My first thought was why would the tech list him as a reference. The next day, we received a call from the applicant. I was on the phone and he asked my secretary if a technician was hired. She asked him to hold and she would ask. He flipped out and started a curse-filled tirade stating he would not work for me if I were the last company in Pittsburgh which was good because I would not have hired him. I read where the person was arrested for a road rage incident. Phew!

Remembering people's names

Dale Carnegie had a saying, *"Remember that a person's name is to that person the sweetest and most important sound in any language."* When you are in sales, you will need to remember names, many of them. Sometimes it is challenging to do that. One of the ways I remember names is to repeat it back to the person and ask if I pronounced it properly. Another way I remember names is to hand-write the name. When I was in my first sales class, my instructor Bob Priganc, taught us to make up some funny twist to remember people's names. He pretended like he was dancing with a pig in front of the whole class. As stupid as that sounds, I still remember it to this day some 30 years later. In case you were wondering how I had people remember my name, I did this skit of a wolf howling at the moon and suddenly having a heart attack. Wohlfarth is a hard name to be tasteful and funny at the same time.

Look for a mix of business

I suggest you consider going after a combination of clients such as service agreements, minor repairs, and installations when planning your sales calls. If you are only looking at replacement or new installs, it may lead to wide swings inside your company. Let us say you are awarded a big installation. You have to manage the installation and when the project is done, you are scrambling looking for new work. Diverse types of business avenues will allow you to plan and manage the business easier.

Reward yourself

One of the ways to motivate you or your sales person is to offer a reward. One of the boiler manufacturers we represent introduced a new high-efficiency boiler model which was about 50% more expensive than the standard efficiency ones. This was at the beginning of the condensing boiler phase and they were difficult to sell. As a way of motivating myself, I promised myself if or when I sold one, I would purchase a large flat screen television. This was back when the flat screen televisions were also new and very expensive. I cut out a couple of pictures of the television I wanted and placed them on the bathroom mirror, desk, and refrigerator. My family thought I was nuts. I finally sold one of the new condensing boilers and the second the commission check cleared, I went to the electronics store and purchased my television.

A contractor used to give away monthly rewards to his technicians who brought in the most extra work. It was something simple like dinner for two, tickets to the ball game, a romantic weekend away in a hotel. He swore it helped to motivate his employees. One of the keys to offering rewards is it has to be a short term contest usually one week to one month. If it goes over that time, the people lose interest.

Setting Goals

Goals will help you attain the things you hope to achieve. Just a couple of warnings. The goals have to be realistic or it will backfire on you. For example, if you set a goal of $3,000,000 in sales the first year with no experience, it may be too high. I would also suggest short and long term goals such as weekly, monthly, yearly, and multi-year goals. The goals should be slightly lofty but achievable. A realistic goal is a 5-15% growth over the previous year. You want goals which will challenge you. Rewards work well when you hit your short term goals and keep you motivated. Keep your goals visible so you think about them daily.

Tracking your goals is also very important as you need to know your progress throughout the year. You do not want to end the year and find out you are way off on your goals. Try to assess your progress monthly. Think of it like the lane departure warnings on vehicles that are there to be sure you stay within the lines. What if you find out you cannot reach your goals? That is the time to evaluate the goals and see whether they were too large or what circumstance caused you to miss your mark.

When setting your goals, hand-write them on paper as it

will embed them inside your subconscious better than typing them with a keyboard. Another creative way I heard is to dictate your goals and listen to them while driving or exercising. I used to like using a word picture for my goal setting. I would write or dictate a scenario where I was on a lush island with my family as a reward for hitting my goals.

In some instances, goals are set by other people. When I worked for a large company, I sold a very sizeable project. The sales process for this sale took several years. The first thing my boss did was to raise my quota as he felt this volume was the new norm. I explained it was a rare sale and it did not matter.

Set up a challenge

I read a story about a new night shift manager for a factory. After he was there a short while, he wrote a figure in chalk on the floor of the number of units his shift completed. The next morning, the daylight shift arrived and asked about the number on the floor. When they found out what it was, they worked hard and made one extra unit than the night shift did. The daylight foreman erased the number and wrote their new number on the floor. This motivated the night shift to outdo the day shift by one unit. The competition increased the productivity of the factory and eventually plateaued but the overall productivity increased.

Keeping the pipeline filled

You have to keep your sales pipeline filled and always talk with new customers. Some companies will work hard at getting a sale and then spend all their time and effort to get the job done and once done, find out there is no work. So the cycle begins again. It wreaks havoc on your cash flow. I recommend using at least one day a week to look for new work, regardless of how busy you are. I found Mondays and Fridays were the worst days to try reaching a sales prospect so I used a Tuesday, Wednesday, or Thursday to contact clients.

Your attitude, not your aptitude, will determine your altitude. Zig Ziglar

About Your Firm

Watch your cash

Cash flow is critical when running a business. My grandmother used to tell me to always save for a rainy day. The rainy day fund is used when sales are slow. Our industry is so cyclical and is said to be either feast or famine. In the off times, you wonder if the phone will ever ring and when you are busy, you wonder if you could clone your workers.

It makes me laugh when I hear people trying to sell me advertising and say "It's a write-off." When I hear someone tell me that, I understand the person knows nothing about business. If you need a write off for your business, take yourself to a beach somewhere and attend a seminar. Guard your money as there are many people out there trying to separate you from your cash.

One of the things I do before buying something for the business is to equate it to billable hours. For example, let us assume I want to purchase a $4,000.00 tool, I will tell myself I have to work 40 hours or a whole week just to pay for the part. I think about the worst job conditions when making my decision. For example, is the part worth me spending a week on a snow-covered roof or inside a filthy basement? If it is, I make the purchase.

Solid Foundation

As with any durable structure, your company will need a solid foundation and this will become apparent as you grow your sales. Think of it as a four-legged stool. Each stool leg will support your business. I believe the four legs are:

1. Take care of the customers.
2. Watch your cash.
3. Learn something new every day.
4. Always look for new business.

Company Name

Naming your firm is something which should take time and consideration. Perform research to help you find the right name. Most technicians look no further than their driver's license to find a name for their company. This may or may not be a good idea if you have a last name like mine. Smucker's Jelly knows their name is not one which makes you think about Jelly and they poke fun of it. Their advertising tag line is *With a name like Smucker's; it has to be good.* You want to have a company name that is easy to remember, easy to pronounce and says what you do.

A friend showed me his new business card and the name of the company was the person's last name followed by Enterprises. When I asked what the Enterprises part meant, he said it was so he could do anything. I explained it would confuse customers. He decided to change it to Heating and Cooling. In my seminars, I use a picture of a London based air conditioning company named Stiff Nipples Air Conditioning Service. I cannot understand why someone would choose a name like that for their business and how long they would be in business. Please select a company name which is not offensive.

Competition

When I attended a sales training class early in my career, I was introduced to one of the most successful insurance representatives in the Pittsburgh area. During the conversation, I asked him how he was so successful in such a competitive market. His answer changed the way I approached my sales. He looked at me and said, *"Ray, There are many people in my industry but I have few competitors."* I loved his answer and asked him to explain it further because I can be a bit thick headed. He smiled and said, *"If you want to be successful, pay attention when you are with a customer. Most sales people are distracted and do not listen to the customer. They are more concerned with trying to get their next sales point out. Always do the small things such as handwritten thank you notes and doing what you promise."*

Lawyers

In your career, you may need a lawyer, and I have a couple of suggestions for you. Find a lawyer who knows your business. Some lawyers in slow times will tell you they know the law and accept a job outside their expertise. An inexperienced lawyer can cost you much more than an experienced one. When I was purchasing my business, I used a friend of the family who was an estate lawyer and not a business lawyer. It was an expensive lesson and I had to change lawyers halfway through the purchase. Ask the lawyer for references of people in a similar business. Ask your accountant to recommend someone.

Donations

We have organizations call us to ask for donations for their fundraising activities. My children went to a small Catholic school and they always had fundraising events to raise money for the school. I like to give away things that compliment my business such as programmable thermostats or a carbon monoxide detector. I will staple my business card to the package. It helps people remember who you are and what you do.

Make it easy to do business with your firm

Our industry is overpopulated with competition and you should make it easy to do business with your firm. When a customer calls, they would like to talk to a real live person and not some machine which makes them choose a mailbox, leave a message, and hope you call them back in a timely fashion. They should have an opportunity to contact a real person immediately. If you are out of the office, either forward your phones to your cell or give your cell phone number to your customers. If someone calls, try to return the call within 10 minutes. A contractor I know has his phones answered 24 hours a day by an employee.

Financing, credit/debit cards

Making it easy to do business with your firm should include taking credit/debit cards and offer to finance. My children never have cash with them and most millennials are the same way. According to GoBankingRates.com, 40% of people say they never use a check. The study also found 61% of the 18-24-year-old people never use a check. According to Gallup, the use of cash has dropped from 36% in 2011 to 24% in 2016. Credit/debit cards will help you grow your business and also help the customer to finance needed repairs or replacement. There are several companies which allow you to use your cell phone or electric tablet to accept and swipe credit cards.

Get your money while there

One of the lessons I learned much too late was to get your money while you are on the service call. This is crucial for residential and optional for commercial. I fell for the *Can you invoice me as I do not have my check book?* excuse several times. People will promise you anything if they have no heat or cooling. Once the system is up and running, your bill goes to the bottom of the payable list. One of the things I learned to help alleviate this is to tell the customer when they call we require payment when the service is complete. Once my tech is there and diagnoses the problem, we give the customer a proposal with the costs of the repair and ask them to sign it. When done, we ask them for either a credit card or check. I like the credit card as we know right away if the funds are available but am not afraid of a check. A credit card processing company charges your company a couple of percent of the sale amount. The customer can dispute the charge and the credit card company can put a hold on the funds. Some people do not like accepting checks in case the check bounces or is declined by the bank because the customer has insufficient funds in the account. That never bothered me as bouncing a check is a felony crime and the owner will pay you the amount of the check rather than risk jail time or criminal proceedings. Another reason to accept a check is there is no charge for using it unlike a credit card.

Cash flow

When you are selling replacement equipment, it can strain your cash flow. You purchase the equipment, wait for the equipment to be installed, and then hope the customer will pay promptly. Since some of my customers did not pay, I now require half up front and the balance when the job is complete. If you get a 50% down payment, it will usually cover the equipment cost.

It is what it is

We sometimes wish people would change their behavior rather than accept them for the way they are. I had a customer who would take between 90-100 days to pay any invoice. It did not matter what I did or how much I called, they paid the invoice on their terms and it frustrated me. I met my dad for lunch and he saw how upset I was. I explained to him about the client and he said, *"It is what it is."* I looked at him and he smiled and said *"No matter what you do, they still pay in 90-100 days. Include the finance charges in your invoice."* I smiled and thanked him. We added 120 days of finance charges onto every invoice from then on. If the customer paid before 120 days, it was a bonus. I was less stressed and happier.

You can't build a reputation on what you are going to do. Henry Ford

Hiring the right tech

Jan Carlzon, CEO of Scandinavian Airlines, turned the airline around during a difficult time for the industry by focusing on what he called Moments of Truth. He felt each interaction of an employee with a customer was a moment of truth for his company. He theorized the customer would judge his firm by the interaction with the employee and that would be their perception of Scandinavian Airlines. Your employees are your Moments of Truth with the customer. If the tech is rude or mean, the customer believes everyone in your firm is the same way. This is why you need to hire the right person. I believe I can teach the employee anything technical but could not teach them how to say please and thank you or honesty.

It's ok to let it go

Service technicians try to repair everything; it's in their DNA. They pride themselves on having the ability to repair anything even if the equipment is beyond the useful life. I like to explain about the life cycle of equipment. I use an analogy of a vehicle with 150,000 miles and the body is rusty, transmission is gone and the engine is leaking oil. I ask the techs if they would put a new transmission in the vehicle. Most would say no and then I would explain how it pertains to old heating and cooling units and whether it makes financial sense to replace a major component such as a compressor.

Paying commissions to technicians

I know many companies will pay sales commissions to their service technicians for sales leads they bring in. I was always leery about paying commission to service technicians as I thought it could lead to selling the customer items they do not need. It is a decision you have to make yourself. I spoke with my techs about it and they said they did not want to be paid a commission as many clients will ask my techs if they are paid commission. They take pride in telling the customer they do not get commission and are recommending the repair because they feel it is the proper thing. We hired a tech who came from a company which paid commissions. He asked me to start doing it and was mad when I told him I would not. He told my service manager about how he would earn extra money by condemning air conditioning compressors which were not defective so he would earn a higher commission. The technician and I parted ways.

Pay commission to the spouse

Service technicians are usually not natural sales persons and many are introverted. If they were good sales people, they would do sales. To generate sales from the technicians, a contractor mailed the commission check to the technician's home so the spouse or partner saw the check. They would ask the technician why they got a commission check and would push the tech for more sales.

As a way of growing it further, he included a monthly total of the commission he paid to every technician inside the envelope. This put more pressure on the techs and he got mixed results and several of the technicians quit so he abandoned sending the ratings.

Fire the customer

I wish I had learned this early in my career. Some customers treat you like a doormat and you do not have to tolerate it. I did my best to please these clients and some will never be happy. In those instances, you may find they are not worth the aggravation they bring. We serviced a building with an old packaged rooftop unit which should have been replaced years earlier. The wiring diagram was missing and all the wires were the same color, black. We told her it should be replaced and she would fight with us about it. She would berate my service technician for taking so long to troubleshoot the unit and then would take 3-4 months to pay her bill. She would make an excuse such as her checkbook was missing. We would have to call to remind her monthly. We informed her we accepted credit cards but she refused. My friend Herb coined a term called the Popeye Moment. Popeye is a fictional cartoon character and would allow everyone to beat him up until he would say his famous line, *"That's alls I can stand cause I can't stand no mores."* He would then eat his spinach and proceed to stand up for himself with his attackers. I reached my Popeye moment with this client and informed her we could no longer do service for her. It felt amazing. Some clients are just not worth the aggravation.

Pita fee

I met a contractor who had a Pita fee for certain customers. Pita stands for Pain in the A**. If the customer were high maintenance or difficult to deal with, he would add a small percentage onto their bills. It made him feel better.

One of the sincerest forms of respect is actually listening to what another has to say. Bryant H. McGill

Appearance

Appearance

I worked with a talented air balancing tech on several jobs. Although he was very smart about his craft, he dressed like he was at a frat house. His typical attire was old stained sweat pants and a long-sleeve tee riddled with holes. The look was completed with old tennis shoes with grass stains. We had lunch a few times while on job sites and he told me he was thinking of leaving the trade because it was filled with "idiots who treated him with no respect." I suggested he would be treated better if he dressed a bit more professional. As soon as I said it, I regretted it. He looked at me as if I had a third eye and informed me he would not wear good clothes to dirty job sites. I acknowledged his concern and simply let it drop. The technician quit the trade.

The first appearance you make with a customer is lasting and difficult to change. You want to be sure you look the part of a professional. Another issue which could affect the first impression is whether your hands are clean. I used to carry heavy duty soap in my truck to be sure my hands were clean before arriving on a sales call. If some of your day is spent doing the work and part is spent making sales, be sure to keep a change of clothes in your vehicle. You do not want to walk into a customer's office with dirty shoes and stained clothes.

Shoes

I was on a job site and saw a computer repair technician walk into a building wearing flip flops. While I guess the only danger is if he dropped the mouse onto his toes, I was still taken aback at a technician wearing something like to a service call. Conversely, if our boots are full of mud from the last job, that shows disrespect to the customer.

Offensive tee shirt

I was on a construction site and saw a worker wearing a tee shirt with a sexual reference. It did not bother me but did one of the office workers. They asked him to leave the job site. Please do not wear a shirt considered offensive on a sales call or allow your technicians to wear one on a service call.

Bad Breath?

I was a smoker and aware it is offensive to most people. To combat the smell, I kept mints or mouthwash in the vehicle. Before going on a sales or service call, I would take a swig of the mouthwash to rid my mouth of the smoke and coffee smell. On one job, an old lady asked, *"Were you drinking? I can smell alcohol."* I said no and explained it was mouthwash. I am not sure she believed me as we never received another call from her.

Grooming

Many sales trainers think facial hair should be gone and cite the CEO'S of most of the Fortune 500 companies and most politicians do not have facial hair. While I had no desire to be a CEO of a Fortune 500 organization or especially a politician, I do like facial hair but believe it should be trimmed and neat. Not many people can sport the Duck Dynasty beards for their job. My son liked to wear a beard when he first started in sales to make him look older when calling on engineers and contractors. He felt it made him more credible.

You smell like beer Mister

I went to lunch with a client and we each had a beer. After lunch, I went to a service call at a school. They were having an issue with a classroom unit ventilator. As I was seated on the floor working on the unit, I said hello to a young student whose desk was next to where I was working. The lad said loudly, *"You smell like beer Mister."* This triggered every student and the teacher to turn around. I excused myself and offered to return when the room was empty. It was embarrassing and I never drank again during the work day. I would refrain from drinking alcohol while on sales calls, even if the customer did. The next sales contact would smell the alcohol and not know whether I had one drink or many.

Uniforms

Wearing clean uniforms provides a clear image of success and knowledge to the customer. Be sure the uniform is clean and not torn or ripped. Once the uniform is stained or torn, it should be replaced. Every once in while you get on a dirty job and I like keeping a jump suit in my truck for those projects. You can purchase them at a big box store and they are disposable. I used to keep a second uniform in my truck in case mine got dirty or wet. I do not want the next customer to think I am a slob. I also keep an extra set of dress clothes in the truck in case I go to a sales call.

Uniform fine print

Be careful signing a contract with a uniform service. My techs hated the uniforms we got from the uniform service company as they were stiff and uncomfortable. We tried to break the contract and they informed us we had a two-year contract and each time we changed something like adding or subtracting an employee, it would restart the time period. It took forever and we finally ended the contract after getting a lawyer involved. Review the fine print on any agreement you enter.

Everything you've always wanted is on the other side of fear. George Addair

Vehicle

When you are driving your vehicle with the company name and telephone number emblazoned on it, be sure it is clean and organized. Some people will judge you by the vehicle you drive. Another issue is the physical condition of your vehicle. I once had a technician with a slight oil leak on his van and it dripped on a customer's driveway. We had to clean all the oil drips from the driveway. If you have any doubts, either park on the street or place something under the vehicle. My techs keep cardboard in their trucks to place under the truck in case of a leak. If the vehicle is in disrepair or rusty, people will assume your work will be the same way.

I once parked in the boss's spot and only planned on being there for a few minutes to drop off a part. When I got in the building, I was busy speaking with the maintenance staff and forgot about the truck. The owner stormed into the boiler room and yelled at me for parking in his spot. I apologized and moved my truck.

Nowhere to hide

When your name is plastered on the side of the truck, it is had to be anonymous. I once had a person call to tell me he was cut off by one of my trucks. I thanked him for letting me know. The truth was I was the person driving our vehicle and the man calling was the aggressive driver and not me. I wanted to tell him what I thought of him and his driving but instead thanked him for his feedback.

Am I being followed?

A less than ethical company in my area was known to follow their competitor's trucks and call on the customers to try to steal them away. I never believed it until one of my techs said it seemed like someone was following him and my client said the company called him to see if they could quote on their business.

Wearing your logo proudly

I like wearing some of my company apparel with our logo when I am out and about after work. I read a study which suggests we meet three new people a day and 80,000 people in our lifetime. I wore the company hats and or jackets while coaching my children and attending their school activities. You will be amazed at how much of a powerful marketing tool it is.

I like to get some nice shirts or hats with our company logo and give them to my employees as a spiff. If they like working for you, they will like wearing the clothes after work.

Offensive Bumper Stickers

Please do not put bumper stickers on your truck or your employee's trucks. Some people get offended at things you consider small. I spoke with a technician who's truck was vandalized because he had a political sticker on his bumper.

We don't develop courage by being happy every day. We develop it by surviving difficult times and challenging adversity. Barbara DeAngelis

Planning for new business

This may not be in your job title but planning is required to deal with the increased business from new sales. It is a not a good idea to pick up new business without the ability to manage it. We were awarded a large contract with a local school district and it taxed our service department. We fell behind on service for our other customers. Also, our response time extended much further out. We lost some longtime customers and it hurt. Ask yourself? How can we handle a 5%, 20%, 50%, or 100% growth?

Another planning consideration is if you are going to put another truck on the road. You have to purchase a truck, shelving, ladder rack, tools, and insurance and easily have thousands of dollars tied up in this decision. In addition, you have to find a technician who will pass a drug and background check, possess suitable customer interaction skills, have a good attitude and work ethic, and be a skilled technician. Good luck with that.

Sales Plan

When I worked for the international company, I had a sales manager who would demand a yearly sales plan. While these are a great idea to help you organize your sales strategy, this sales manager would be over the top. His sales plan was more than 200 pages long and would take almost a month to complete. I brought it up to him and mentioned this was taking away from my actual selling. Not a good thing to say. He threatened me with being written up for not completing the plan. Welcome to corporate America. It seems like some of these companies make money in spite of their best efforts. When I started my own company, my sales plan was two pages long. It included the following:

Sales Goals This is the number of sales I would like to bring in and what percentage was service, service agreements, and equipment replacement.

Where should I focus my efforts? I limit this to the three markets where I want to increase market share. An example of this may be residential within a certain area or zip code, retail stores within 2 miles of my business location, and motels within 10 miles. How do I know which ones to try? I look at the existing customer base and use those for pushing my sales effort. In the above example, I see how many calls are within a certain zip code. I would focus on that area so my techs are not driving all over the place. That is the same for choosing retail clients within a certain mileage of your location. You can respond faster. Calculate your existing market share in each area and try to increase it.

How will I get the sales? You talk about the actions you will take to get the sales in this section. This may range from how many cold calls you will make to how many sales appointments you will have a week.

What road blocks will I face? List anything from what happens if we lose technicians to what happens if the weather is mild and service calls drop. I would not spend much time on this but give it some thought in case.

What steps will I take to get better? This is where you discuss what training you would like to accomplish this year. Is it visiting the ASHRAE convention? attending sales training seminars? or purchasing sales training videos?

Take a client to lunch

If you want to increase sales, take a customer to lunch and ask them how they would suggest you increase sales. You will be amazed at some of the responses. Some may offer to call some friends in different facilities or suggest certain organizations where you could attend. One of the questions I like asking customers is, *If you were me, who would call on?* They may have a colleague who is looking for HVAC service.

What is your niche?

If you want to be successful, I urge you to specialize in one area of the trade. The person who attempts to do everything will usually earn much less than the specialist. I read a survey of earnings for doctors in a 2016 Fortune magazine and saw the difference in earnings and it was stunning. For example, the doctor doing family medicine earned an average salary of $206,000 per year. While the salary is substantial, consider the average salary for an Orthopedics Doctor was $443,000 or 115% more. It pays to specialize.

There are many areas where you could specialize in the trade. My area of expertise is boilers. I met an owner of a company who specializes in rooftop units. He has a sheet metal shop to make Adapta curbs and his own crane and can install a rooftop unit the next day. One company I know can seal ductwork and balance the airflow to make the entire house more efficient and comfortable.

How to find your niche?

I fell into my niche by accident. I was working for a control company and they sent me to school to learn boilers. I fell in love with the idea of boilers after listening to the passion the boiler class teachers had for them. How do you discover your niche? You want to find an area where you are passionate is the first step. For example, What type of equipment do you like working on? The second is to see if there is a market for your niche. One friend specializes in old steam boilers. It is a limited market and certainly not a growth market but he can charge whatever he wants.

Well done is better than well said. Benjamin Franklin

One big company or lots of small ones?

When I started my business, I read a story about a company which made high-quality porch swings and distributed them locally in their area. A large retailer approached the owner and wanted to sell his swings regionally. He agreed and his sales started to grow. The customer then approached him about selling the porch swings nationally. The owner got a business loan and started a factory to meet the demand. The demand for porch swings by the large customer monopolized his business and the smaller customers soon fell away. Shortly after, the big customer started making demands. They wanted him to cut his prices or they would look somewhere else. No sooner had the owner cut his costs, the large customer wanted another price cut. The owner reviewed his finances and informed the customer he could not lower his costs any further. The customer abruptly canceled the orders and the owner had a brand new factory with no customers. The owner had to close his doors. If I can impart one bit of advice for you to heed, do not allow one customer to have all your business. Instead, try to have small sales to many customers.

Commercial or Residential

I prefer commercial customers as they are easier to access. We can call the customer and usually be able to go there the same day. It allows us to use it as fill in work. The residential home is sometimes difficult to schedule a service call as the owners typically both work and have to plan a day from work to allow you access. *Whatever you do, you have to make sure you make it there when the customer takes a day off work.* The cash flow for a commercial business is slower as you have to invoice the customer and wait to be paid. This may take anywhere from 30-60 days. The cash flow is much better on residential as you can get paid the day of your service call.

Another comparison of commercial and residential is the hours. I see a trend in residential companies that offer extended normal service hours where they are charging the same rate all the way to 7:00 pm. That works well for larger companies with more technicians but small companies like mine are hurt by that as my techs start at 8:00 am and are tired by 4:30. I believe the technicians are a valuable asset and never want to wear them out.

How some business is better than others

Think about what type of business and customer you want. We made the mistake of bidding on the maintenance contract for several highway facility buildings with the state. It was one of the biggest blunders in my life. We cut our rates to get the work so we were already losing money.

Also, these people were a nightmare to please. Since the buildings were manned 24 hours a day, we would get calls in the middle of the night complaining about everything from a squeak to an odd smell. I think they were lonely. You know that smell when a heater starts for the first time each season? They would call about that smell the first week the heat was started and we would have to respond within two hours. When we found a defective part, no matter how expensive, we had to call the customer's main office to get approval to replace it. It seemed like every time we called, the person who could make the decision was in a meeting and we would have to wait for him to return the call. The time my tech spent there waiting could not be billable. Neither was the travel time to get the part. I could not wait for the contract to end. Think about what type of customer you want to bring to your company. It is sometimes better to let your competition deal with the painful ones.

Would you like to double your profits?

Our industry has an average profit percentage of 2-3% which is too narrow for the type of work we do. If you want to double your profits, consider raising your prices by 2-3%. This will drop down to your bottom line and your profit is doubled to 4-6%. This is not great but it is getting better. Most clients would not balk at a 2-3% increase and it makes a huge difference to your company bottom line.

You cannot cross the sea by standing and staring at the water. Rabindranath Tagore

What is the difference between markup and margin?

Some think these terms are interchangeable and they are not. The margin is the sales price minus the parts cost. If a part sells for $100.00 and the cost is $70.00, the margin on the part is $30.00 or 30%. Markup is the amount added to the cost of the item to arrive at the sales price. If we have that same $70.00 part and our markup is 30%, the sales price for the part is $91.00. The markup required to achieve 30% margin would be 42.9%. Margins will allow higher profits than markups of the same amount. The following is a comparison of margins versus markups.

To get this margin:	Use this markup:
10%	11.1%
20%	25.0%
30%	42.9%
40%	80.0%
50%	100.0%

The formula for calculating markup is:

Markup = Margin divided by Cost

Higher margin vs. low bid

Try to look for higher margin work rather than low bid type of work. Consider your time is finite and if you can make a net profit of $50, $75, or $100 per hour rather than $40, life will be much better. The key to getting higher margin work is to look at selling more expensive equipment. For instance, a builder's grade on off single stage condensing furnace may cost you $900. If you mark it up 35%, you will make a gross profit of $315.00 for installing the furnace. Consider a variable speed multi-stage furnace with a cost of $2,100.00. If you marked this up 35%, your gross profit would be $735.00 or 233% higher gross profit for roughly the same amount of labor. Try to sell the higher margin equipment. Since the equipment is more efficient, it is a win-win for you and the customer.

Can your firm do the work?

A friend and customer of mine left his job as a maintenance supervisor for a mall and took a job at a food processing facility. He wanted us to bid on doing the service for the walk-in coolers and freezers at the facility. We had to decline the opportunity as it was not our expertise. I had this vision of a cooler filled with frozen ravioli ruined because we could not get the unit fixed in a timely manner. My company's expertise is boilers and comfort HVAC. Always work on growing your core market and expertise.

What is the cost per hour?

Our industry has an average billing percentage of 50-70%. This means with a typical 40 hour work week, we will be able to invoice between 20-28 hours per week. The other lost time is due to call backs, sick days, holidays, vacation days, traffic jams, or other causes. So for calculation sake, let us assume we get 24 hours billable labor per week. It equals 96 hours billable per month. This is key to remember.

New Service Van Monthly Payment	$600.00
Monthly Insurance Cost	$200.00
Fuel	$200.00
Repairs, tires, etc.	$100.00
Total per month	$1,100.00

Your truck cost per hour is $1,100 divided by 96 hours = $11.46 per hour based on 24 billable hours per week. If you can get 30 billable hours per week, the cost per hour for the truck drops to $9.17.

The same calculation can be estimated for every cost you have. Another expensive cost is hospitalization. For example, we pay $1,300.00 per month for a family hospitalization plan for our employees. The cost per hour is $13.54 based on 24 billable hours per week. It drops to $10.83 per hour or 20% lower if we can get more 30 billable hours per week. It pays to stay on top of billable hours.

Chicken or the Egg

There is an old saying, Which comes first; the chicken or the egg? Do you purchase and outfit a truck, hire a technician, and then grow your business? If you do, you could dig yourself a hole which is difficult to exit. The other option is to grow the business and then hire a new technician. It could be tough also as you could have angry customers if you cannot satisfactorily service them. I believe the correct answer is a bit of both. When I worked with a large company, my manager would only let us hire a new technician when the existing technicians were working 20% overtime. For example, the normal work week is 40 hours and we would only be able to hire a new technician if they worked 8 hours overtime a week. He felt the 20% overtime allowed us to have enough work for an additional tech and we could grow the business to find more work for the new tech.

What type of customer?

When looking to grow your business, consider carefully the type of customer you want as you will be married to them for a while. For example, I dislike working on the equipment inside a bar. My family had a bar when I was growing up and my job was to clean the spilled and stale beer and I hated it. I would not want to call on bar owners for their business.

You have to consider the cash flow ramifications of the new business type as well. For example, if you work with a general contractor, they sometimes take a while to pay their invoices. If you bring on a new employee, you have to invest lots of money in outfitting the new tech and their truck and have to wait 30-90 days for your money. This will quickly eat away your savings. You will need to speak to your bank about a line of credit and you never want to do this when you need money. There is an old saying, *The bank will never give you money when you need it, they only give it to you when you do not need it.*

How many sales should a sales person generate?

This is a difficult question to answer as it depends on the profit margins you have in your business. A sales person should generate at least two times their salary in gross margins just to break even. I like to see them generate close to 3-4 times their salary in gross margins. Let us assume your firm does primarily repair and replacement work in residential and would typically have 35-40% markup on the labor and materials. If the average sale is $5,000, each sale will generate between $1,297.00 and $1,429.00 in markup. If your salesperson is earning $50,000 per year, they would need to bring in between 70-77 projects per year at $5,000.00 each or a little over one project per week, every week and this is just to break even. If you want to earn money with your sales person, they need to sell between 100 and 140 projects per year. It equals between almost two and three projects per week. That is a lot of replacement sales.

One of the harsh realities I learned was you have to be an asset for your firm. The minute you become a liability, you are expendable. If your firm is paying you a salary, they expect you to not only to earn enough to pay your salary and benefits but you have to turn a profit. If you cannot do that, you are in danger of losing your job. In case you are wondering, a rule of thumb is benefits and taxes are about 25-30% of a sales person's salary.

What's the margin with you?

Some sales people are consumed with sales volume and want to be paid for that. Sales volume never tells you the whole story. You should focus on the gross margin of the sales and reward the person for that. For example, if the person who is paid a salary of $50,000 per year sells $200,000 of sales with a 20% markup, it will bring $40,000 to the business. Congratulations, you just lost $10,000. If the same sales person sold $200,000 with a 30% markup, you now went from losing $10,000 to making $10,000. Huge difference my friend.

You miss 100% of the shots you don't take. Wayne Gretzky

Pricing considerations

When pricing a proposal, I like to use odd numbers which make it difficult to divide and find the price per unit. For example, you purchase a six pack of Guinness Beer and the cost is $12.00. You would know easily the price per bottle is $2.00. Now, if we purchase the same six pack at another store for $10.49, the price per bottle ($1.75) is not as easy to calculate. When I sell my modular boilers, I make it an amount hard to divide.

Another idea I like to do is to round down the price. For example, I will use $488.00 instead of $500 in my pricing as it looks much less. I never liked the idea of using just under the whole number like stores do. They sell their items for $499.99. I believe it looks less trustworthy.

Should the sales person be involved with customer retainage?

You have to decide what works for your business. I think a sales person is beneficial to your company if they are bringing in new customers. If your sales person is seeing the existing customers, who is bringing in new customers? The manager or owner should be the one in charge of customer retainage. A sales person paid for customer retainage will merely add to the overhead costs.

A sales person for a large mechanical contractor started to visit customers and take them to dinner and lunch. The boss did not think there was anything wrong as he viewed the increased expense reports. The manager was surprised when the salesperson gave his notice of resignation. A few weeks later, the manager started getting letters from customers canceling their service agreements. The sales person started his own company and stole more than half the customers and almost 80% of the business. Lawyers were involved and it took several years for the original company to regain their sales volume. The owner always handled the customers after that.

Should they get a commission for sales to your existing customers?

This is often a sore spot with many sales professionals. They believe once they get a customer, it is theirs for life. I understand the appeal to do this as sales to new customers is difficult. If you are going to let your sales people handle the customer retainage and pay them commission, I would not pay the same commission rate as the one for new customers. It should be much less. You want to reward them for getting new work.

You can only control actions not results

One of my sales managers was a wise man who felt sales quotas were hard to control. He was an engineer and believed if he could control actions, the sales would follow.

Rather than focus on the sales quota, he had certain tasks I had to complete each day. He wanted me to do ten cold calls per day. He felt I should have 3-4 appointments with every ten cold calls. If I did not get the appointments, he worked with me on my cold calling technique. Once I got the appointments, he expected a closing ratio of 50%. If my closing ratio dropped, we worked on my closing and sales presentation. It was an ingenious formula and able to be managed. It led to high margin sales.

Growing your business the wrong way

There was a local company with a growth philosophy to gobble up the sales people for any competitor. They felt by waving money at the sales person; they will bring over their customers. I interviewed with the company and they did offer me a substantial raise if I came to work for them. The only thing they wanted was a list of the customers we serviced and the sales volume. I respectfully declined their offer as it would have made me feel dirty and sleazy. I found out it was the right decision. They hired sales people and tried selling all their customers and when done, they let the sales person go. Be wary of companies like that. Do not sell your soul.

Visit a client

If your business is slow, visit your existing customers. They will appreciate it and you will find some work. Explain how things are slow and ask if they have any projects or problem areas.

Get your company involved with a home warranty or insurance company?

Each week or so, we get calls asking if we want to work with a home warranty or insurance company. I have not had much success with these organizations. We tried a couple and it has been a nightmare. One wanted us to double our insurance limits and keep our labor rates the same. Another took five months to pay. I know some companies work with these firms with success but I would not do it again. I understand when you first start your business you have to take anything to generate cash flow but be wary of these. I had a friend who was owed thousands of dollars and the company kept stalling him. After 3 ½ months, he called and asked to set up an appointment and wanted to get his payment. During the meeting, the contact for the company offered him a check for half of what was owed. My friend balked at the offer and the man looked him in the eye and said: *"Either take this check or you can take me to court and spend all your money in legal fees."* My friend took the check. He read later the company went bankrupt and the other vendors only got $0.10 on the dollar for what they were owed.

Always check credit

We had a general contractor who wanted to do work with our firm. I was lazy and did not check the credit references on their credit application. If we had, we would have found out the company had a terrible reputation and did not like to pay their bills. We eventually got our money after six months, many phones calls and two letters from our lawyer. I check every credit application now.

Conversely, I met the sales person for a large control company and he told of a new policy in their firm. Each prospective customer had to complete a credit application before the sales person was allowed to visit. I was amazed and asked how it went over with his customers and he smiled and said *"not so good."*

Cash is king

When running your business, be sure to watch your cash flow. Be careful of going with less than reputable customers. We once sold a furnace to a *"real estate investor"* and it took us six months to get the money. We had to call each month for payment. I explained we took credit cards and he did not have one. He called a few months later and asked if we would bid on a new furnace at another of his homes and we informed him he had to pay up front. Do not be your customer's bank. Another drawback to dealing with slow payers is most sales people have the unenviable job of collecting these past due bills and it takes away from your sales effort.

Guard your cash

I had an office manager who seemed like a great employee. She had a good attitude, always on time, great with the customers, and dedicated. She paid the bills and balanced the checkbook. I started to offer her more responsibility. The only thing I did not do was to allow her to sign my checks. One day, she called off and I gathered the mail. I saw the bank statement and a little voice in the back of my head suggested I open it. This was back before online banking and the banks still sent you copies of all your cancelled checks. I started to leaf through the checks and found one made out to my office manager for an odd amount of several hundred dollars. I flipped through a few more and found another made out to her. A sick feeling overcame me. In all, I found four checks made out to her totaling several thousand dollars. I opened the check book and found the reason for the odd amounts. She chose invoices for different vendors and wrote the exact amount of the invoice. Instead of making the check payable to the vendor, she made it payable to herself. It was brilliant on her part. If a vendor called and asked where the payment was, I could look in the check book and tell them the check number and date. The thing that also got to me was it was my signature on the checks. I could not imagine me signing a check made payable to her and I do not sign blank checks.

One of my techs found something in the garbage can in the shop to explain it. It had page after page of her signing my name on blank paper. She had forged my name and did it well. In total, she took almost $10,000 in cash and about $7,000.00 in credit card charges. I felt so violated. Keep your check book under lock and key and balance your own check book. She was arrested and had to pay back all the money she stole.

Don't lower your prices

This lesson has cost me thousands of dollars in lost revenue. When times are tough and business is slow, you will be tempted to lower your prices to get new work. I caution you against doing it for a couple of reasons. The first is the customer who purchases only on price are usually the most demanding customers. Secondly, if your existing customers find out you are selling your service for a lower rate, they will be mad and expect you to lower your price to them. One of my competitors used to have labor rates all over the board and I used it against them. I would tell their customers who were paying the higher rate about the lower rate paid by other customers. For example, If I knew certain regular customers of my competitor paid a rate of $80.00 per hour while the client I was meeting paid over $100.00 per hour. I would let them know about the lower rate. Now, if this customer is paying the higher rate, it will make them question whether they are being treated fairly. It opens the door for me.

RAY WOHLFARTH

The other consideration with different rates is the lower rate somehow becomes the new standard rate. The bosses who approve or pay the invoices will question why the rates changed. What I like to offer instead is a discount. I may tell the customer we have a 5% discount for the month of April. The invoice still shows our traditional rate but we have a discount lower.

Mileage charges

We charge a mileage fee for our service calls and it is higher than the amount allowed by the IRS or Internal Revenue Service. Occasionally, the customer will complain and want the same price as the IRS allows. I will explain how that is for a person using his or her personal vehicle. I describe our trucks are filled with $40,000 in parts and tools to get the heating and cooling repaired in a timely manner.

Fall down seven times. Get up eight. Proverb

Finding New Customers

Using Social Media

Some consultants recommend using social media like Facebook, Twitter, and Linked-In to help grow your business. While allowing access to your social media seems like a good idea, it could backfire if you have posts offensive to some of your clients. If you have political statements different than what the customer feels, it may cause your reputation harm.

Barter?

My brother sold custom tee shirts and belonged to a barter company. He would provide a certain amount of goods and then was able to purchase something he needed or wanted from another member and no cash was exchanged. I investigated it and opted not to participate. I did not want to have my money tied up and unavailable. Another version of barter is when the customer offers you their products if you provide something for them. If you service bars, these people will love doing this. Do not do it! My friend replaced a rooftop unit for a bar and agreed to accept free food and beer instead of a payment. Since he liked food and really enjoyed beer, he thought this was an excellent idea. Everything worked well for the first six months until he went to the place and found it shuttered. My friend lost several thousand dollars.

Searching through old files

One of the things I like to do is to go through the old customer files and contact them to see if they are interested in speaking with us about doing their service again. We have had customers leave due to our prices and then realized the new cheap company is not a good value. I like to stay in contact and ask them if we could earn their business back again. In some instances, they are embarrassed to contact you again since they left. I try to make it easy to do business again and would never gloat about their decision. Many customers want to return once they find the new company is not what they thought. They just need to be asked.

Go to the tallest building and look around

I used to work for an international air conditioning firm before I purchased my business. My sales manager went with me to a sales call at a tall building in my home town. As we stood on the roof of the building, he looked out and instructed me to contact every building with an air conditioning component on the roof. You could do the same thing in your town.

You just can't beat a person who never gives up. Babe Ruth

Read the newspaper

My sales territory for boilers includes three states and I like to read the local newspapers in the cities I visit. It gives me something to talk about when I visit my customer. It allows me to seem like a local to them and interested in them. I will scan the headlines a couple of times a week and book mark any stories I think would be interesting to my customers.

Get help from the government

The government can be a good friend and supply you with a great source of leads if you know where to look. I specialize in boiler sales and service and almost every state keeps a list of every commercial boiler installed in the state. The list includes everything you need to know to contact the prospect:

- Building location
- Building owner
- Equipment type
- Equipment manufacturer
- Age of Equipment
- Who to contact

Once you get this list, you will be amazed how many boilers are in your state. You could load the list into a spreadsheet or database and sort it by age and would retire before you get to the end of the list. Another list which will give you some pertinent information about potential clients is to get the list of elevators in your state. I realize your firm most likely does not service elevators but what is in every building with an elevator? If you answered air conditioners, you would be right. You could also sort the list by zip code and call on these clients when you are in the area.

Sunshine laws

So how do you get this treasure trove of information? I would suggest you look on the Department of Labor and Industry website for your state. Some states keep the entire database on-line and you could download it to your computer. It is usually already in a database which is accessed by a spreadsheet program like Microsoft Excel, Access, or Mac Numbers. At the time of writing this, my state does not have their list on line. To get a copy of the list, write a letter to the state requesting it. Since it is public knowledge, they have to provide it to you under the Sunshine Law. They may charge you a small handling fee. I would include the words Under the Sunshine Laws; I am requesting a list of elevator or boiler installations. My request was sent to the Bureau of Labor and Industry, Boiler Division.

Fortune favors the bold. Virgil

Residential

If we are replacing a residential heating or cooling unit, I like to contact the neighbors and let them know about our company. One of the ways to do it is by looking at the local real estate website. Many local municipalities will keep records of each home. The letter will explain what we are doing for their neighbor and ask if we could discuss their system.

Borrow from others

A homework assignment is to look at all the mail you get at home and the office and see what jumps out at you. What catches your attention? Then think about how you can incorporate it into your marketing and sales plan.

Join a professional organization

When you are trying to attract new customers, you might want to focus on certain niche markets. It is much easier and cost-effective to focus on one group rather than spread your resources. I am involved in the local college and university market so I belong to the local APPA group and speak to the members at their meetings. I chose that market as I knew they would always have money and pay their bills. Look for organizations in the markets you want to penetrate. In my area, there are professional organizations for school district maintenance departments, school district business managers, commercial building, and hospital facilities. The local chamber of commerce is also a good source for leads.

Cold Calls

Most sales professionals will cringe when I suggest cold calls. They envision the customer waiting for them with a medieval torture device. I used to like doing cold calls and had fun with them. For one thing, most sales people never do a cold call. They talk about it but never actually do it. I have found most people are very receptive as it provides some relief from their daily life. In reality, it is not that bad. You will not have to do cold calls your entire career. Once you are successful, people will find their way to you.

Warm calls

Warm calls are calls when your friend or current customer recommends you to the new client. The closing percentage is much higher and the sale process is much less stressful. You could ask your existing customer if they know people in their position at other organizations and if they would introduce you to them.

Use a Mirror

A sales trainer recommended placing a mirror on the desk to be sure you smiled when calling clients. He believed people could tell by your voice if you were smiling and felt you will be more successful if you are smiling.

Should you go or should you stay?

The average cost of a customer contact is $33.11 for a telephone sales call and $276.48 for a field sales sale call according to Salesforcetraining.com. If you are selling an item or service valued at $200 or less, it would not be financially prudent to visit the customer and the sales effort should be via a phone.

How to get your letter opened

If you want to get your letter opened by the customer, try hand addressing the envelope. Think about your own mail. You get all sorts of junk email and subconsciously throw them in the trash. If you see a handwritten letter, it will cause your brain to stop and take notice. That is all you want to do, separate yourself from the junk mail.

Send a dollar

A salesman I know includes a dollar bill in a letter to

prospective clients telling them this is the first of many dollars his company will save them. This is too hokey for my tastes but my friend said it gets results. A financial planner I know has a business card that looks like a $1,000,000 bill and he sends it to his clients and tells them he could help them reach a million dollars.

How many calls a day?

I used to work for a boss who wanted me to make twenty cold calls a day. It was incredibly difficult as you are traveling all over the county to see people. Also, it was challenging to do the follow-up and still do the cold calls. The follow-up is just as important or more than the cold calls. We compromised on ten cold calls a day, four days a week. That left me to do my paperwork on Fridays. What I found is with every ten cold calls, I was able to get 3-4 appointments. I would end up with about one sale for every ten cold calls. Ten calls are still quite a bit of work but doable.

What is each customer worth?

I found a site, ContractorSelling.com, which has some incredible numbers for the costs and value of a residential customer. They contend the average cost to obtain a residential customer is between $250-$450. While it seems high, they argue the average customer over a 30-year business arrangement can bring in $42,700.00 in profits. They believe an average homeowner will require the following components over a 30-year period:

- 2 furnaces
- 3 air conditioners or heat pumps
- 3 humidifiers
- 2 air cleaners
- 30 yearly service agreements
- 15 Service calls

I believe the value of a commercial customer is easily double the amount and most likely much more.

What is each cold call worth?

Sometimes it is hard to stay motivated when doing sales cold calls. It can sap your confidence when you hear that many no's. One of the things I did was to break down the value of each cold call. My sales to closing ratio was 10% on cold calls which meant for every ten cold calls, I had one sale. If you consider the above-estimated profit over a 30-year period to be $42,700 and divide it by ten, each cold call is worth $4,270.00. Think of it this way; if the customer is mean and nasty, it is still worth over $4,000. When I walk away from the call, I mentally thank the jerk for giving me $4,000. I know it's a mind game and it helps me to keep going.

The customer will not return your call

My friend, Tim calls on purchasing agents inside industrial complexes and it is almost impossible to reach someone if they do not want to talk. His idea was to purchase a business book and he would include a business card and a note inside the book such as *"I thought you would like this."* Each time he did it, the customer called him later and thanked him for the book and invited him in for a meeting.

If I want to try reaching someone and they never return calls, I consider it a challenge. I will write a letter to the person and the letter will do a couple of things. It will try to motivate the customer to call me or at least answer the phone when I call. The letter will also include a couple of quick sales bullets about what we have done for other customers. It will tell the customer I only need a few minutes face to face. Inside the envelope, I may include a few of the reviews we got from other people in the same industry. I will conclude the letter by saying I will be calling on a certain day and time to see if we could schedule a meeting. Does it work? Sometimes it did and sometimes it did not but that is the nature of sales.

Whether you think that you can, or that you can't, you are usually right. - Henry Ford

Offer two times

A way to get an appointment with a potential customer is to offer two different times. It makes it easier for the customer and does not seem like you are pressuring them. For example, I would say this:

Ms. Client, I am available on Tuesday at 9:00 am or Thursday at 1:15 pm. Which time works better for you? This is not asking for an appointment. It is asking which time is better. See the difference?

Offer unique times

I met a sales person who would set his appointments for odd times such as 1:12 pm or 9:34 am. It seemed phony for me but he thought it worked well for him. I did not like it as the odd times did not fit well into my planner.

Donate to the charity the customer likes

A sales person for a company was trying to get in to see the owner of a large company without success. He read how the owner and his wife hosted a fundraiser for a local charity. The sales person donated to the charity in the man's name and sent it to him along with his business card. The owner called to thank him and they did business together.

You have 15 minutes

I love getting a meeting where the customer tells me I have 15 minutes to give him my sales pitch. You can handle this in two ways. You can use humor or you could speak rapidly. I prefer using humor. One of the ways I would handle this is like this:

Client: I will give you 15 minutes to tell me what you know.

Me: Oh That's easy. I can tell you everything I know in 10 minutes.

Now the client will usually smile and allow you more time. My friend had this happen and he walked in with a kitchen timer. The customer asked him what was going on. My friend told him *"I wanted to make sure I didn't go past 15 minutes."* The customer laughed and told him to proceed.

Getting through voice mail gate keeper

In this age of connectivity, why is it we cannot talk to anyone on the phone? We get pushed into voicemail and they never call us back. One of the ways of bypassing the voicemail trap is to dial a different extension. For example, if the customer number ends with 4560, try adding a single digit to the number and dial 4561. This may bypass the voicemail box and connect you to a real live person. I always press the 0 to see if I can speak with a live person when I call a customer.

Using a 3x5 card

I like to have fun doing cold calls and use a trick Rosemary Launikonis showed me. I write down the name of the decision-maker in a building or company on a 3" x 5" card and then walk into the building and ask to see the person. The receptionist will inevitably ask why I am there. I will hold the card up and say I am not sure but perhaps there are some areas of the building which are either too hot or cold. Are there any areas like that in the building? 99% of the time, the receptionist with an electric heater on will readily agree. If it does not work I may ask, Have they been having problems with the heat or cooling breaking down? I will then say I was told by the office to hand-deliver this information. In many instances, the receptionist will contact the person.

Finding out who to contact in a facility

In many companies, there is a sentinel who guards the boss against unwanted interruptions and that person is usually the receptionist. I have found if you call and ask who to be connected to the maintenance department, they will not connect you with the key person in most instances. One of the things I like to do is call the main number and tell them I have to mail information to the key person and ask who I should put as the contact. In most instances, the receptionist will gladly inform me who I need to contact. They will usually tell you the name of the person if you say you are only mailing the information. I guess it is less intrusive than calling the person. I can either wait and call a few days later and ask for the person by name or stop by with my 3x5 card.

Look at the log in sheet

Whether I am going to a sales call with a new client or visiting an existing one, I always look at the visitor log in sheet. I want to see who was there. If it is a new prospect and I see several recent visits by a service technician for my competitor, I can assume they are having heating or cooling issues. I will keep this in the back of mind for when I am doing my sales call. I might ask if they were currently having any comfort issues. If it is an existing client, I may look to see if any of my competitors were visiting the building. I use the excuse I want to be sure my techs were signing in. While visiting a client, I saw my competitor visited twice within a week. It set off warning bells inside my head. When I met with the customer, I happened to mention I saw my competitor visited the customer. The customer was a bit taken aback and I asked, *"Is there something we are doing wrong? I would like to know if we can fix it."* The customer informed me his foreman did not like the technician assigned to the building and requested he consider another service provider. I asked if we could meet the foreman and find the issue he had with us. During the meeting, the foreman told us the previous company gave hats to the workers and we did not give him caps for his workers. When he asked about hats, he felt my technician was rude to him. I knew the tech was not normally rude and perhaps he was having a bad day. I apologized to the man and my contact and asked if we could resolve this problem. He agreed and we brought hats and donuts in the next morning for him and his crew. They liked us after that. Remember your difficult customer is someone else's hot prospect.

Don't stalk the client

I was speaking with a customer one day and he asked me if I knew a certain person who worked for a competitor. I told him I had never met him. He then said the person called him ten times one day while he was out of the office. *"What kind of person does that? Am I his only lead? I will never do business with his company."* I asked how he knew the person called that many times and he said his phone would tell him all the missed calls. If I offered advice to the sales person, I would tell him to limit the calls to once every day or two.

When to find the person

Most of the executives you want to contact do not work standard banking hours. If I want to contact them, I will call early in the morning around 7 am or after working hours typically after 5 pm. If you are contacting a maintenance department, many of these people start work very early in the morning. If I am doing a cold sales call on this type of prospect, I show up at 7 am. This is usually before the receptionist is there.

Use company website

The company website for the client you are looking to do business with usually list the key personnel. I look on their website to see whom to contact. The site may even include the telephone extension and email address.

Trade shows

I like going to trade shows and many sales people have the misconception the trade show is an automatic way to close sales. In many instances, it is simply a way to meet potential customers. I like to use it to see who to contact and perhaps have a flyer sheet listing my customers who are in the same industry. For example, if I am at a trade show for school districts, I would list all the educational customers I now service. If you get an opportunity to teach a class at the trade show, I urge you to do it. It gives you a chance to display your expertise and meet more decision makers in one day than you can in a month of sales calls.

High expectations are the key to everything. Sam Walton

Talk to vendors that sell different products

During the slow times of the trade show, I like to introduce myself to the other vendors at the trade show. I will tell them what we do and leave a flyer or card. When the show is over, I will send out a quick note saying how I enjoyed meeting them and look forward to sharing leads some time. This works out in two ways for my company. Some of my new friends will remember me and offer my name to their customers. I now have significantly increased my sales presence. The other benefit is I have someone I could recommend to my customer if they are in the market for their product. It endears me to the customer as the "Go To" person to call to get answers. If my client needs their service or product, I send an email to the vendor and include my customer in the email. I will tell the vendor about my customer and ask them to take good care of my customer. The vendor and customer are both appreciative. The vendor will look to repay the lead.

Get Involved with the organizations

If you decide to market your company to a certain market, try getting involved in their professional association. Volunteer to help and they will readily accept. I like to volunteer to be on the greeting table as I get to meet all the people when they arrive and they get to know me. I used to plan the golf outings for an organization of hospital maintenance supervisors and spoke in front of the group several times a year.

Meet with big companies

Some contractors have made a nice niche for themselves by offering their services to the local branch of an international company such as Carrier, Honeywell, Johnson Controls, Siemens, or Trane. In many instances, the local branch labor rate is much higher than yours. In addition, their technicians are usually skilled at service and maintenance but not at installation. Your firm could fill that void. Just be warned some of the large companies take forever to pay their invoices. We sold boilers to one company and it took three months to pay and we were on the expedited list. We do the boiler startup and services for several of the large companies because they do not have the in-house expertise.

Tip clubs

Tip clubs are lead-generating organizations which meet once a week. It is a group of business professionals from diverse business categories. This is one of the advantages to using a tip club. For instance, you would probably be the only heating cooling company in the group. Each attendee will usually address the group and tell them what type of customer they are hoping to meet. Some of the tips clubs require you bring a couple leads each week. I felt

LESSONS LEARNED SELLING HVAC SERVICE

uncomfortable recommending some of these people to my clients. I was always afraid it would reflect on me if the person I suggested were a jerk.

Want more sales? How about hiring more salespeople?

I organized my own tip club comprised of people who would call on the same type of customers I did. For example, my tip club included a water treatment company, plumber, roofer, general contractor, and electrician. These were all people I worked with and felt comfortable recommending to my customer. We would meet a couple of times a month for breakfast.

Look for open windows

An old-timer gave me this bit of advice when looking for places to contact. If the building has open windows, it usually means the people inside are uncomfortable and the owner's fuel costs are too high.

Keep a recorder by you all the time

I used to keep a tape recorder with me all the time and now I just use my cell phone to keep notes and remind me of events. If I see a building I believe would be a good prospect, I will record where it is and the contact number. I will also use the recorder to remember ideas I had when driving. My friend uses the virtual audio assistant on his phone to send himself emails and notes.

Talk to anyone within 3 feet

A sales trainer once told me this strategy and I am not sure if it helped grow the business but it certainly embarrassed my children which made me smile.

Hand out your business card

When we did residential heating and cooling, I would hand my business card out to everyone. I would include a business card with the tip for a restaurant server, hair stylist, and even include a card in all the bills we paid at both home and the business.

Who was that again?

When someone hands me a business card, I like to write on the card the date I received it , where I received it, and a note on the card about the person. This will help me remember the person and what we spoke about during the meeting. I only file them in the business card file after I send the person a quick email with a note about what we spoke of and thanking the person for the time. It will make you much more memorable.

Employee discount

We offered a discount to the employees of our commercial customers for their personal residence. They already knew our work and our employees and it worked well.

Fundraising

Another way of growing your business is to offer it as a fundraising arm for a local organization. Every organization from the PTA to the local school sports teams are looking for ways to raise funds. This is an excellent idea as a portion of whatever you sell will go to the organization. For example, if you charge $99 for a furnace or air conditioning check, you could offer a percentage back to the organization. I did this for my son's soccer team and it was a win-win deal for both of us. Another item we did was to donate a furnace preventive maintenance call as part of their auction. Our company was listed in the event flyer as a sponsor and we got several new customers from it.

Offer your expertise to the media

If you let the local media and news reporters know your expertise, they may contact you for an expert opinion. For example, you could be the expert they could contact about why a furnace check is a safety benefit or the dangers of carbon monoxide poisoning. Another good site is www.helpareporter.com. It allows you to register your expertise and consult for reporters.

Write articles

A way of getting yourself known as the local expert is to write an article and submit it to the local newspapers. It could be something from why check your furnace or air conditioning unit to common carbon monoxide dangers. This works on two fronts. The first is you get your name and company to all the people who read the local magazine. The second is to make copies of the articles and hand them out to clients. It will make you look like the expert. An unwritten rule is never send the same article to different magazines. If they find out, they will never accept an article again.

Use technology to sell

I like to use a carbon monoxide detector inside a boiler room or basement to look for dangerous conditions. On a residential call, I will offer to test the water heater and oven for carbon monoxide.

Perimeter marketing

I was taught a trick by a friend who called it perimeter marketing. While installing a residential furnace or air conditioning unit, he would visit the houses immediately surrounding the home where they are working. He would drop a flyer at each home and offer an exclusive sale on a new unit or service call. Since you are doing the work there already, it adds credibility. It only takes a few minutes to drop them off.

Flyers

Many companies make up flyers to hand out in neighborhoods where they want to gain market share. The key to a successful flyer is the opening statement. You only have a few seconds to grab the clients attention before they throw it into the trash. My favorite opening of a sales flyer is to ask questions we commonly hear in the sales calls but have fun trying different options. You can have a couple of different styles and see which one works for you. As I said, I like using questions to try grabbing their attention and changing it according to the season. Include pictures on your flyers to grab attention. Some of the best pictures to catch attention are a pet, baby, or an old ac unit. My opening would be something like this:

Cooling Season

1 Does your home feel clammy? I would have a picture of a

clam. It sounds funny but works.

2 Does your home feel like something from Goldilocks and the Three Little Bears?

Upstairs too warm?

Basement too cool?

First floor just right?

3 Does your home have three different climate zones?

Upstairs is rain forest?

Basement is North Pole?

First floor is perfect?

4 Is your heating system ready to retire?

Heating Season

1 Does your home feel drafty?

2 Do you have to turn up the television volume when the furnace starts?

3 Do you wake with a stuffy nose every morning?

If you answered yes to any of these questions, we could help.

I would then include a story about how my company helped

the customer resolve the above problems. Have fun with these and continue to tweak it. Remember you have to show in only a few seconds how you can help make their home more comfortable or safer. To close out the flyer, you have to include a call to action such as asking them to call or email you to start making their home more comfortable. I have used flyers for commercial buildings as well. Most tenants in a strip mall are required to perform HVAC maintenance on the units for their space so I will drop off flyers to the stores.

Use an electric combustible gas detector.

A small percentage of all gas trains have a leaking component or fitting and the gas detector will help you find the leaks. I like using the electronic leak detector as it legitimizes us to the owner. In many instances, the customer will watch as you leak test the gas piping. This will increase sales and make your customer's home or building safer. I used to use soap bubbles painted onto the joints but found it would drip onto the floor and I had to clean the floor when done. I once had a service technician who used his cigarette lighter as a leak detector. The customer said he thought he could smell gas in the boiler room and told our tech. The technician walked in with his lighter and ran it next to the gas train. This did not go over well and the customer told him to leave. The customer told me to never send the tech there again. The weird part is the tech had a leak detector inside his van but did not want to go out to his truck.

A contractor I know purchases stick-on carbon monoxide detectors with his company name and telephone number imprinted on them. He hands them out to customers. They have a large dot on them which will change color in the presence of carbon monoxide. The only drawback was these only last a few months and I would not want the family to trust these with their safety. I urge my clients to purchase a carbon monoxide detector from the store. They are relatively inexpensive. One contractor gives away a carbon monoxide detector to all new customers. He advocates for carbon monoxide safety and feels this helps with his brand.

Air Flow Hood

Without a doubt, one of the best investments we ever made was an airflow hood. This tool, although expensive, paid for itself many times over. These are the same air capture hoods used by air balancing companies to test and adjust the airflow to a building. Once you learn how to use it, you will generate lots of work and satisfied customers. On one commercial three-story building, the first floor was a nightmare for the building management firm. They had customers calling all day long complaining about the temperatures. Our initial investigation showed a temperature difference of 15-degrees F from one side of the floor to the other. We looked above the drop ceiling and saw flex duct kinked and missing or blown off. We gave the customer a price for fixing the problem. We guaranteed a 3-degree difference in temperature from one side of the floor to the other. The customer agreed and we fixed all the ductwork and used our air hood to balance the air flow. When done, the office space was within the three degrees we promised. The tenants loved it and the complaints dropped. We got a contract for the other two floors and now have much of their work. When you use a flow hood, you will most likely be the only company the customer has seen using one and you will stand out.

If your firm specializes in residential, this can pay you benefits as well. It seems like many homes with forced air heat find a substantial temperature difference between the top floor and the first floor.

Give us your worst building

One of the challenges I offer to potential customers is to let me look at their worst building. If I cannot fix it, I will not bother them again. This challenge works amazingly well. It had also bit us in the past when we were unable to fix the problem but my thoughts were we did not have the service before and this simply was our shot at earning their business. I would not do this for free as anything free has no value. You could offer a discount.

Thermometer

My previous employer had custom thermometers with a magnetic back which would show the temperature in the space. He had an arrow pointing to the comfort zone which was between 68 and 72 degrees F. The thermostat would instruct the customer to call our company if the temperature was outside the comfort zone. They were very popular.

Look for the hidden gem

I was asked to give a training session to the custodians of a school district in my area. They did not want to be there and made it abundantly clear by talking and calling to each other while I was speaking. I found myself getting angry and tried to focus on the small percentage of the people paying attention. When done, a young man came up to me and apologized for his co-workers. He asked me several questions and took my card. A few years later, he called and introduced himself again and told me he was responsible for the maintenance of several buildings. He asked me to see him and we serviced his buildings for several years.

No one can make you feel inferior without your consent.
Eleanor Roosevelt

Using Seminars to Sell

Give a speech

Many community organizations look for speakers at their meetings. I enjoy giving a talk as it lets me reach many new prospective customers and allows me to look like the expert. Everyone would rather work with the expert.

Glossophobia is the fear of public speaking and is the leading phobia for most people, rating higher than the fear of death, spiders, or heights. It is a great adrenaline rush and remember, you know lots more than the people in the audience. If you would like to hone your public speaking skills, consider attending a local Toastmasters club. This organization will give you tips to become a better speaker. If you find you enjoy public speaking, you could attend a local National Speakers Association club. This organization is for people looking to make a living speaking.

Even though I have been speaking publically for many years, I still get nervous at the beginning of the talk. I use a PowerPoint presentation to help me stay on track. It helps me stay focused. If you service the residential market, you could speak to organizations such as the Chamber of Commerce, Rotary, local library, or Kiwanis Clubs. If you do commercial service, you could talk to the local real estate investors club, Chamber of Commerce, or industry organizations.

Teach a class

The local community colleges or libraries like to have classes and you could teach one directed to homeowners. It could be anything from what homeowners should know about furnaces or air conditioning to seasonal maintenance including furnaces, air conditioners, and humidifiers. The key to this is the attendees are qualified leads for your firm and you come across as the expert. Be sure to hand out a flyer or business card to the attendees.

An additional venue to teach a class is to arrange to hold the class in one of your client's facilities. It helps the customer and helps you if other facilities you currently do not service are invited. We had a college host a training seminar and they collected the money for the event. They were able to generate some money for their maintenance department. Another way to use training is to meet with local real estate agent offices and offer to teach a class. You will be the expert and they will remember you when their customer needs HVAC service or replacement.

Training classes

I like to offer free training for my commercial clients. It accomplishes a couple things. It cements my relationship with the customer. It also allows us to talk about new safety upgrades to their system. Another benefit is you get to know other employees in the company. This is valuable in case your contact leaves. It also allows you to service the home HVAC systems for the employees.

Lunch and learns

These are used for engineering firms or the maintenance departments of large school districts or universities. You agree to provide lunch for the employees and have an hour to teach a class about your product or service. The class should not be a one hour commercial for your firm. The best use is to show the attendees how you resolved common problems with your service or expertise.

Free seminar Part 2

A friend in the water treatment area told me he rented a hotel room to give a free talk about water treatment. He invited all the hospitals and school district maintenance staffs. On the day of the talk, only about half the people showed up and he was hurt. He asked my advice and I told him the only problem was the seminar was free. If it is free, people think it has no value. On his next seminar, he charged a small fee. His attendance was much better.

Do or do not. There is no try. Yoda

Free evaluation

When I am doing a talk for a local organization, I may offer the attendees a free evaluation of the boiler room. For example, I offered to do a free walk through of one of their more problematic boiler rooms for a local college. I found several things and made my recommendations. Over the next few years, we sold over 40 new boilers to the facility. I look around the boiler room to see any code violations or dangerous situations such as the combustion air openings, and evidence of leaks or soot.

How to give a speech

The first thing to remember is the speech should never be a commercial for you or your company. It will backfire and the attendees will have negative feelings about you and your company. I have our logo in the corner of the slides and a large logo on the first and last slide. I like to discuss issues commonly found in their industry. If it is a residential group, I may talk about why the second floor is so much warmer or colder than the first. A typical talk I give is the dangers inside a home; cracked furnace heat exchanger, carbon monoxide, fuel leaks, safety controls. The talk will include how regular service can make it safer. When speaking to a commercial group, the talk may be on safety, energy reductions, green buildings, or how to reduce comfort complaints. I like to pass out cards to the attendees for future contacts. The more talks you do, the better you become at giving them.

What if your competitor is in the audience?

This was one of my worst fears when I first started to give seminars. I had this vision of the person trying to make me look bad to all the attendees. In all my years, I only had one difficult person. I acknowledged his presence and told the audience, We have a representative from XYZ company so you have two experts and I am sure he would be glad to answer your questions after the class. It seemed to shut him up.

If your competition is giving a seminar, I will do my best to attend. I will approach the speaker before the seminar and ask if I could attend and promise to not say a word. I just inform the speaker I was hoping to learn from them. Most people have no problem with you attending. It will give you a clue on how they go to market.

Building your presentation slide show

When I first started giving training seminars in the industry, I used an overhead projector and it was a nightmare. Today I use a projector and PowerPoint, it is so much easier. The following are some tips I suggest when building your PowerPoint presentation.

Use pictures – There is a saying, A picture is worth 1,000

words and it is perfect advice for a PowerPoint slideshow. Pictures are much more effective than words.

Use stories – I like to use job site stories when showing a picture. It helps you relate to the crowd.

Avoid sentences – The fewer words you use on each slide, the better. What I like to do is use key words on the slide and embellish the meaning when I talk. Use no more than five lines of text on each slide.

Font Size - When compiling your slide show, use font sized so the person in the rear of the room can read it.

DON'T READ THE SLIDE - Sorry for yelling but please do not be the person who turns his or her back to the audience and reads the slides. It will insult the attendee and even anger them.

No special effects – Please do not use the special effects on the PowerPoint slides. Many people new to PowerPoint think it is cute to have the text come bouncing in from the side or use awkward noises such as a typewriter keyboard sound. One of the fastest ways to lose your audience is to use these effects.

Proper Placement - Plan your seminar so you do not reach across your body and turn your back to the audience. If you are right-handed and use your pointer in that hand, stand to right of the screen as you face the screen. If you are left-handed, stand on the other side of the screen.

Start on time, End on time – This is crucial, some people will be offended if you wait to start for the stragglers. You also want to quit on time. It is inconsiderate to infringe on the time for the next presenter and the attendees may want to go to the next seminar and it could be in a different room.

Allow time for questions – I like to end with a few minutes left for questions. If it is an hour-long seminar, I will end it about 5-7 minutes early so I can field some questions.

Repeat the questions – When an audience member asks a question, I like to repeat the question so everyone in the room will hear what the person asked. The audience gets frustrated if they hear you answering a question without knowing what was asked.

Have laptop screen facing you – When you are speaking and your back is to the screen, position the laptop screen so you can see what is displayed on the screen behind you without turning around.

Get a laser pointer - I like using a laser pointer as my pointer as it can be seen with you standing across the room. If you use your hand or fingers, you are blocking access to the people seated on the side of the room where you are standing. In between speaking engagements, you can use the laser pointer to play with your dog or cat. Laser pointers do not work on television screens so you may need to use something else to point.

Use your own projector – As you start to do more seminars,

I urge you to consider purchasing your own projector. I have had instances where I was unable to use the projector at the facility and my laptop would not connect to their projector.

Don't look at your watch - I have a digital timer I use for when I am talking. It is visible only to me while I speak. It will allow me to know when to wrap up the talk.

Talk slower – If you practice your speech, it will go much faster when you are speaking in front of an audience than during the practice. Either add some more slides or talk slower.

Get audience involved – Your seminar will go much better if you get the audience involved. I use humor in my talks and try to find people to engage. You can ask questions and get a show of hands for the answer.

A sale is not a sale unless you get a purchase order, you complete the work, the customer gives you a check and the check clears the bank. Ken Launikonis

Advertising

You could go broke advertising for your business and have to be very careful with the advertising budget. I have tried advertising in church bulletins, newspapers, industry newsletters, and other areas. I am not sure what worked and what did not. I think people get inundated with advertisements and yours gets lost. I would recommend focusing on one or two venues.

Perhaps I am a grumpy old man but I have a hard time believing the sales people who tell me how successful advertising with their magazine, bulletin, or newspaper will be for my company. I have tried almost every type of advertising arena and found limited results. I believe advertising works if you have the budget of General Motors or McDonalds, but I never got the results I was hoping for with it. I believe the only person who benefits from advertising is the advertising sales person. I now look at advertising in places where it will benefit my customers. For example, If I do work for a church, I will sometimes place an advertisement in the church bulletin. I decide on advertising on a case by case basis. I like to advertise in the sport or school programs where my children attended school because I knew it was going to a good cause. I have been in this industry for several decades and can count on one hand the calls I got from an ad I placed.

How did you get our name?

One of the best ways to test your advertising methods is to ask every new customer, How did you get our name? This will tell you where they found your name. You can decide if it was from your advertising efforts or something else.

Target customer type

When you have a small organization, you most likely do not have unlimited funds for advertising and marketing. If that is the case, I would focus your attention on two to three different markets and try to increase market share in those markets. How do you know which market to choose? I would look at your existing customer list and categorize it by market type. I would then think about which clients you have a good relationship with and focus on that market. You are going to use the customer to leverage your market share. For example, if you look at your client base and see you currently service two colleges and have a good relationship with them, I would focus my attention on contacting all the local colleges and universities. Another way to increase the market share in that area is to see if there is a local organization where you can meet the contacts. For instance, I like working with colleges and universities. There is a local chapter of APPA of which I am a member. It is much easier talking with the potential customers in a more relaxed setting like a meeting. Many of the markets have organizations where you can meet some of the decision makers. The following are some of the organizations which may hold meetings close to you.

ASHRAE - This organization is comprised of the engineers who design heating and air conditioning systems and the equipment manufacturers and representatives.

BOMA - This is an organization of commercial building facility professionals.

NAPE - The National Association of Power Engineers is an organization of stationary engineers who work in the facilities department for large buildings.

Other organizations – I would check and see if other local organizations are in your area where you could meet potential clients. I belonged to a real estate investors group, school facility managers, craft brewers and distillers, school business managers, and hospital facility engineers. A good way to find groups in your area is to check the Meetup groups.

Balance

When getting involved with the organizations, it can be very time-consuming. I was involved with several organizations and missed many important events at home and with my children. Try to use balance so you can spend time with your family. Also, these time hogs can pull you away from your job and affect your performance.

Advertising specialties

You can get your name and telephone number printed on almost anything. I have run the gambit of purchasing customized pens, pencils, screwdrivers, thermometers, coffee cups, calendars, travel cups, and rulers. The things I had the most success with was a refrigerator magnet with a thermometer on it for residential customers and coffee mugs and screwdrivers for the commercial and residential customers. My brother who sells advertising specialties has a belief to give people their favorite coffee mug and they will always remember you.

There are no secrets to success. It is the result of preparation, hard work, and learning from failure. Colin Powell

Driving sales from service

The service department can help you generate sales because the customer already likes you. It could lead to growth at higher margins. Once you start thinking about sales, you will see a change come over you while doing service. You will start seeing opportunities you may have overlooked before while on a service call. I urge my techs to take 10-15 minutes after the service call to look around the boiler room for any safety or efficiency opportunities. For example, look for leaking pipes, dirty evaporator or condenser coils, or missing insulation.

Make it cleaner than when you arrived

I like to clean the equipment I work on as I believe it shows pride in my work. I was called to a problem job with a burner. I found and repaired the problem and wiped the burner down when I was done. It took me about 5 minutes. When the owner of the building walked in, he looked at the burner and asked what happened. I told him I just wiped it down and he smiled. He told me no one had ever done it before and we eventually got to do the service on all of his buildings.

Show respect for the client

I met a potential customer on a job site where he wanted to discuss the purchase of one of my boilers. As I walked into the building, I saw a black ink looking stain on the beige hallway carpet which led from the basement to the rear door. I knew what it was as soon as I saw it. It was the black sludge found inside old boilers. There were foot prints from workers who accidentally stepped into the dirty water and tracked it. I knew from experience the carpet was ruined. When I met with the customer, I mentioned the spill and he shook his head and told me the contractor who did it was fired from the job. If we are replacing a furnace or boiler, I like to line the pathway from the basement to the outside with a carpet liner. These are available at the big-box stores. I like to hose out the old boiler and enclose the furnace or boiler inside the thick contractor bags to limit the chance of spilling something on the carpet or floor. When I am doing a residential service call, I like to use the paper booties over my work shoes. I believe these show respect for the client.

Ask your existing clients

One of the things I found is the customer you have now knows their peers in other facilities. For example, if you service a school district and you want to work with others, you could ask your existing client how to get to know other school districts. He or she may be able to introduce you to someone in another school district or tell you about a local organization where you may meet others. *I like working with schools and universities as they pay their invoices promptly and are a safe credit risk.*

If you do sell something to the person the customer recommended, be sure to let your customer know. I would send a handwritten thank you note and included a small gift card to a Starbucks or someplace like that.

Train techs what to look for

The service techs can be a valuable asset for your sales. They may balk at being thought of as a salesperson but you need to explain they are not sales people, they are providing solutions. They have to be trained on what to look for while on a job site. I will explain about the life expectancy of equipment and when it is time to look at a replacement system. I will also review the building codes and show them how to look for opportunities. I also review the benefits of system upgrades like a programmable thermostat, humidifier, UV light, or better filtration.

Service agreements

Service agreements are very important to growing a business as it allows several things to happen:

Pull through - When selling a basic service agreement, there is a rule of thumb which suggests you will get an average of three times the sales volume of the service agreement in extra work from that customer. For example, if you have a basic service agreement of two visits per year at the cost of $500, the additional sales found during the service calls or emergency calls should total an additional $1,500 per year on average.

More leeway when pricing a replacement - You may not have to be the low bidder when a replacement system is required. In many instances, you will be the only bidder. They would rather work with you.

Takes client off the street - If the client chooses a service agreement with your firm, it takes the client out of the feeding frenzy which is our market.

Increases the value of your company - If and when you wish to sell your business, one of the sale points is recurring income. The recurring income from service agreements can increase the value of the business.

Even Cash Flow - A service agreement helps you get money throughout the year even when the sales are low. On commercial buildings, I like monthly invoices to even cash flow.

Helps manage labor - A service agreement can help you allocate the man hours of your company and allows you to have someplace to send service techs when the business is slow.

Status Report

One of the ways to increase commercial sales to existing clients is to use a status report. I take pictures of the equipment and document the age and condition of the equipment. I will use the ASHRAE life expectancy in my report. For example, if the customer has a 14-year old furnace, I will note in my report the age of the unit and show the customer the ASHRAE study which suggests a 15-year life for a furnace. My customers have thanked me for the report as it allows them to budget for the replacement of the equipment.

Hot potato

There was a game I played as a child called Hot Potato. It consisted of a group of people in a circle who would pass a tennis ball, an orange, or a potato to each other until the music stops. The person holding the ball when the music stopped was out of the game. When you present a letter to the customer detailing a safety issue with their equipment, it is like that game. No one wants to be the person that was aware of a safety issue and did nothing when they were warned. As a result, it cannot stay on their desk. The person has to do something with the letter and they usually pass it up the company and you get the repair.

Completing the Service Report

When filling in the service report, more words are better. My service technician went to a job site and replaced a flame safeguard on the boiler. The technician wrote on the service report, No Heat. Fixed it. My client called and jokingly said we charged him $500 per word. I had to rewrite the service report and send him a copy of everything the technician did. Remember the service report you or your employees complete may be viewed by people further up the organization who know nothing about what we do but have to approve the invoice for payment. Give them a short story.

Include Model and Serial Numbers

On every service report, I urge you to include the model and serial number of each piece of equipment you service. It will document the equipment you worked on and protect your firm.

Let customer save face

Several times in my career I have been called to a job site where the custodian or maintenance person turned off the boiler and forgot about it. They were so embarrassed when we found the cause of the service call and asked me not to tell the boss. I was in a bit of a quandary as I do not want to lie to my customer. I allowed the custodian to save face and suggested we check the safety controls while we were there.

Customer Thermometers

One of the most frustrating things I have seen on job sites is when a customer uses a free promotional thermometer to second-guess you. These promotional thermostats may be off by several degrees. When I walk in with my professional thermometer, the customer may argue and say their thermometer is more accurate. I like using a digital thermometer as people believe it to be more accurate than an analog one.

Wrong Temps

We replaced a defective thermostat which had a temperature reading off by several degrees. We set the new thermostat for the same temperature as the old one. The client said the new thermostat was defective. We had to show the customer with our digital thermometer the temperature reading was correct and suggested he set it to a temperature which felt comfortable to him. By listening to the customer, we were able to sell a replacement system the next year.

It's not 70 in here

Many customers do not understand the temperature difference between the setpoint and the actual temperature. According to industry standards, a system is considered balanced when the temperature is within 3-degrees of the setpoint. That could equal a 6-degree temperature swing. It is almost impossible to have the same temperature in every room unless each room has individual controls.

Using codes to increase sales

One of your best friends is the building codes your municipality follows. It allows you to quote the code and bring the customers equipment up to meet the current code. I like to train my techs on the new code changes so they can look for opportunities with the customers.

You can also use the codes to increase service sales. For example, my state follows ASME CSD1 which states in section CM 130 the owner has to have

- maintenance performed on their automatic boiler
- record the findings in a log or service invoice
- Any defects shall be brought to the attention of the owner and corrected immediately

I would urge you to keep a copy of the current codes with you or your techs and be able to show it to the customer.

How are you doing with the existing clients?

This is one of my favorite ideas for generating sales. I will mail a letter to every one of my customers essentially asking how we are doing. The letter is on our letterhead and includes the name, address, and company of the client. The letter has a few lines for comments. I would not ask about the cost of our service as everyone responds they think we are expensive. I do not want to get that thought in their mind. There is also a self-addressed stamped envelope for the customer to return the survey to us. I keep copies of the questionnaires in a binder and use it for sales calls. This does two things: It lets us know how we are doing with our existing customers and uncover any who are not satisfied with us. The second benefit is for new clients. If someone is looking to do business with us, I let them look at the binder. I keep the returned letters enclosed in a plastic sheet. The clients will look at the names and their comments. What do you do with not so complimentary reviews? I include them and tell the customer about the ones where the customer was not happy. I explained what we did to resolve the problem. It works great. The following is a typical letter we send to our customers.

Success is never final. Failure is never fatal. It is courage that counts. Winston Churchill

Ms. Customer
Acme Real Estate
Address

Dear Ms. Customer,

We really appreciate your business and hope you can help us provide better service to you and our other customers. Please answer the following questions and return to us with the self-addressed and stamped envelope. Thanks for your help.

Please rate the following:

Response Time: Excellent___ Good___ Fair___ Poor___

Tech Expertise: Excellent___ Good___ Fair___ Poor___

Communication: Excellent___ Good___ Fair___ Poor___

Was problem resolved quickly? Yes____ No____

Comments_____

_____

Very truly yours

Ray Wohlfarth

I didn't know you did that

I visited a customer's job site to look at the packaged rooftop unit we serviced and saw they had a brand new boiler installed in the basement. I asked the customer why he didn't call us for the boiler. His reply shocked me, *I didn't know you did that.* I just assumed the customer knew everything we sold or serviced. Please do not make that mistake with your customer. Every time I see that boiler, it reminds me of the lesson. Let your client know of all your offerings.

Ask the techs

I like to ask my technicians who they think I should contact about replacement equipment. They know which units are in poor condition. If I sell it, I will give a gift card or a new tool to the technician for their help.

Spiffs

Try offering sales spiffs to the technicians to grow certain segments. It could be something like a $25 gift card for every new humidifier or air cleaner sold.

Don't let what you cannot do interfere with what you can do. John R. Wooden

Marketing

Thanksgiving Day

I had an accountant who would mail hand-written thank-you notes to his customers around Thanksgiving Day. The card would tell me, the customer, how thankful he was to earn my business. It was something different and I liked the approach.

Fundraising Event

A friend of mine was the co-sponsor of a fundraising run to benefit a local charity. The event would garner publicity locally and the local news stations would cover the event. He was interviewed each year and would wear something which showed his company name and logo. It was an inexpensive marketing idea.

Newsletter

I like to send a monthly newsletter to my clients and prospects. Mine is a digital newsletter. Newsletters are a tricky way of marketing as you have to be careful not to make it look like a sales piece. If you do, the customer will simply throw it away, opt out, or ignore it. I like to include success stories in our newsletters. I use

Constantcontact.com for my newsletters. It allows me to see who opened the newsletter and which links they clicked on.

Google News

Google News is a good site as it allows me to scan through the news stories about subjects of interest to my customers or me. You can add keywords to filter your search and it works great. I will reference these articles in either my newsletter or to send them to customers.

Let the others know

If you get a new client, let the others in the industry know. For example, if you start doing business with a real estate management firm, I would contact every other real estate management firm in my marketing area and let them know about the new client. No one wants to be the Beta Tester and it opens the door for them to start using your firm.

Contests

I once had a contest where I would send out a playing card each month with a sales sheet about one of the products we sold. It was a poker game and at the end of seven months, I would see who had the best hand and gave them a free trip. You can get a fair price on a trip if you arrange it several months in advance. It was very successful and the manufacturers we represented pitched in and gave us money for the trip. All in all, it cost us nothing, and we got lots of free publicity.

Door Stickers

I went inside a large commercial building and saw a sticker on the revolving door saying the building's energy was being monitored and controlled by a local HVAC company. I am not sure how many building owners would allow it but I thought it was clever.

Linked In

This is a business site where you can contact other business professionals. It is a good place to do some marketing of your company and yourself. You can use it to post successful projects and discuss some of the challenges of each one. I would include links to stories which you are passionate about. This may include energy efficiency, green buildings, or carbon monoxide dangers.

Facebook

You can use Facebook to promote your company as well. Be

careful to not to make it a sales ad. I would use subtle sales tactics such as success stories or announcing a new customer.

Twitter

Twitter is another site where you can post short updates or tidbits about your company.

On-Line suggestions

Be careful posting commercials about yourself or your company. Most will be ignored. I prefer posting links to articles about boilers, carbon monoxide safety, energy savings, and healthy buildings. I find those articles by performing searches on-line and posting links. Another way of extending your marketing arm is to visit the sites of industries where you want to increase market share. For instance, if you have expertise in carbon monoxide dangers, you could offer to answer questions from their members.

Inviting a customer to the game

My boss used to have two season tickets to the local professional football team. He wanted me to use them as a sales tool and invite a different customer to the games. It did not turn out to be such a great idea for us. First of all, my family and I felt a bit of resentment as I was taking time away from my family on the weekend. It was also very uncomfortable at the game as I had to be careful what I said or did as the guest was a client and not a friend. We had some get visibly intoxicated and I had to drive him home. One customer expected to go to every game and was disappointed when I told him I had to take other customers. It turned out to be a marketing nightmare as the client was angry with our company. We also offered the tickets to the customer so they could take whoever they wanted. This was much better. I suggested to the boss we give the tickets to a deserving employee, best decision we ever made. It motivated the employees and got us away from entertaining.

Yard sign

Many residential installers have yard signs they place in the front yard of any replacement project. It lets the neighbors know your company is doing the work. Try leaving the yard sign in there for a week or so. Some companies even have a box where they place flyers or business cards on the yard sign in case a neighbor would like some more information.

Opportunity does not knock. It presents itself when you beat down the door. Kyle Chandler

Sales Presentation

Attention Getting

When you start a sales presentation, you need something to catch the attention of the customer. I like using questions to grab their attention. Think about what you want to say as it will set the tone of your sales presentation. I try to gear my question according to the person's industry I am meeting. If I were meeting with a commercial real estate management company, I might say something like this:

"Ms. Client, we have helped our customers to slash comfort complaints, lower utilities, or reduce repair costs. Are any of these an issue at your facility?"

Try to think about what is important to the customer.

Some sales people like using a shocking statement or statistics to grab the customer's attention. The following are some examples:

Shocking statement: *Did you know there are over 400 deaths each year from carbon monoxide poisoning in the US?*

Statistic: *3 out of 10 heating systems are in poor condition and dangerous.* **This is fictional and simply a way to show you how this opening statement works. My lawyer made me include this.**

Humor: *Did you know the brain works optimally at 40 degrees F but unfortunately employees do not.*

Offer heat loss

Offer a heat loss calculation when quoting a furnace or air conditioning replacement. It will allow you to properly size the system and separate you from the pack. Explain what happens when a system is oversized or undersized. If you do not have the software, there are several places on line where you can get a manual sheet to do the heat loss. An important part of this process is to get the customer involved in the process. I would have them hold the end of the tape measure or assist in adding the numbers. It allows them to own the process. Instead of me doing the heat loss, it was *We*.

Ranking the clients

Try ranking your sales prospects to see how often to contact them. If the customer has a large potential sales volume, you want to contact them more often than the customer worth a few hundred dollars. In my area of boilers, the customer with boilers over 20-years old would be a better sales call than one with two-year-old boilers. I would contact the one with the older boilers more often than the one with the new boilers.

Use scripts

Try developing scripts for cold calls, appointments, and

follow-up appointments. They will help you stay on point and focused. You will need to practice to make them so you sound non-scripted. They will help you with your sales.

Checklist

My friend Tim uses a checklist for his sales calls to be sure he covers all the points he wanted to cover. His reasoning is airline pilots and surgeons use checklists and it helped him with his sales.

Organize sales calls by area

A database with all your customers and prospects is a cost effective way to target sales by location. For example, if you have an appointment, you could look in your database and see what other prospects may be in the area making you more productive. You could even schedule appointments by area for each day. Mondays could be all people in the north, Tuesday all clients in the south, etc.

Change the conditions

We were asked to bid on the replacement of a five-ton condensing unit and A coil. We were bidder #3 on this project and I knew we would be more expensive than the other two companies. While there, I saw the furnace had only a side return duct and remembered this furnace required either two side returns or a bottom return to work correctly. I showed the owner the return and he gave us the sale even though we were more money because he felt we knew more than the others.

You have been underbid

In our industry, there are some companies known as the "low bid" companies. I have found the customers who only consider the price will switch to another company if they are slightly less than yours. I would rather sell my expertise and value rather than compete on price. Another rule of mine is, the customer who considers only price will be a demanding customer.

In some instances, we do have to compete on price. I will sit down with the customer and break down our hours. For example, if we are charging the customer $800 per year for two service calls, that equals $400 per call. I tell the customer the average service call is 4 hours. If my competitor is charging half of our costs, I would ask the customer how he or she thinks they could do that and offer to cut our service from twice a year to once. Either they are not doing the service we are or are using less skilled technicians. Be careful of this if your firm was only there two hours on each visit in the past as the customer will know.

Reduce it to the ridiculous

Let us assume you have been underbid by $1,000 on a furnace. A way to minimize the price difference is to reduce the price difference over the life of the equipment. For instance, I would say, Ms. Customer, our price is $1,000 more than the competitor. I agree it seems like a substantial amount. Consider this: The furnace will last for an average of 15 years. The cost differential is less than $0.18 per day. Do you agree the benefits to your family are worth less than a quarter a day? Now, the difference between you and the other company is less than a quarter. You could even hand a quarter to the customer to drive home your point.

What would you like me to omit?

When looking at cutting costs, I like to list all the items we are doing and show the list to the customer and ask what we should omit. It adds value to what you do. Another thing we found is some competitors will give a lower price and hope to skate by without doing maintenance because they know they can do it for a few years on a well-maintained system.

Am I boring you?

When my children were babies and sleep was at a premium, I went to a sales call after a rough night of teething babies.

During the sales call, I yawned a couple of times. The customer stopped in the middle of what he was saying and asked, *"Am I boring you?"* I apologized and explained what was going on. Since he had young ones at home, he understood and suggested we reschedule the meeting. I thanked him for his understanding. Stay well rested so you can be sharp in front of the customer.

My family bought me a Watch for Father's Day and I liked it as it allowed me to see my emails and messages by rolling my wrist. During a meeting, I was deluged with emails and kept checking my watch. The customer asked if I needed to be somewhere else. I apologized and decided to never wear the watch while on a sales call. A salesperson I know leaves his phone in the car during a sales call and will not answer any calls, texts, or emails. He feels it allows him to concentrate on the customer.

Do not hand anything to the client

When you pass something to the client, you have lost control of your sales call. Try to resist handing anything to the customer until you have to. They will read it and not pay attention to you. If you do hand something to the client, wait until they are done reading it before you restart your sales process. You want a customer who will concentrate on you and your presentation.

One of the things I like to do is to mail something to the

client after the sales call. For example, if we were discussing a subject during the sales call, I may make copies of an article or website which would support what we spoke of and send it to the customer. One customer was telling me about how the deer in the neighborhood ate his flowers. I found an article on how to deer proof the yard. I wrote a short note on the article saying, *"I hope this helps with your deer problem"* and mailed it to the client. I got a call a few days later from the client thanking me for the article and saying we hope to do business together. It had nothing to do with HVAC but it helped me sell the customer.

Always Lookee Eye

One of my favorite movies was Karate Kid about a young man who was bullied and an older man who taught him self-defense. In the movie, the mentor, Mr. Miyagi, gave this advice to Daniel, his student, Always Lookee Eye. It meant to always look the opponent in the eye. As a sales person, you need to learn to do that with your customer. The customer will never trust you and your sales will suffer if you cannot make eye contact with the client. It does not have to be a staring contest but you need to look into their eyes so they can read whether to trust you. I had a tough time doing this in the beginning and was taught to look just above the customer's eyes at his or her forehead. After doing that, I was able to start looking them in their eyes.

Using Analogies

These are great sales tools and should be part of your sales presentation arsenal. Vehicles work great for analogies as everyone can relate to those. I will give you one I use for doing heating system maintenance. In my locale, we have an average of 4,000 heating hours per year. It means the home or building will need heat 4,000 hours per average heating season. I will compare it to a vehicle operating at an average speed of 25 miles per hour. At that speed, the vehicle would rack up 100,000 miles in the first winter alone. Most people would never operate their vehicle for 100,000 miles without having it checked or the oil changed. The clients can relate to this.

Guess how much is in their bank account

Some salespeople will play a game and guess how much money the customer should spend on their new system. They either base it on their own personal budget or on the car in the driveway. It is a dangerous habit and will cost the sales person money. I remember reading about Sam Walton, founder of Walmart and one of the wealthiest men in the world, used to drive around in an old Ford pickup. He could have afforded any system I was selling.

We were asked to look at a replacement hydronic boiler project in a home. The owner already had two other quotes and we were there to simply provide another. I could have opted to price a cheap boiler and hope I was cheap enough to get the job. It would have been a sale of about two to three thousand dollars. Instead, I walked through the house and asked questions. I did not have a presupposed budget for the project. As we were walking through the living room, I spotted one of the largest cast iron radiators ever and it was right in the middle of the floor. I asked the owners about the radiator and they both winced and told me it was such an eyesore. They then said there was nothing which could be done. I smiled and shook my head. We removed the first-floor radiators and installed an amazing heating and cooling system complete with underfloor radiant heating, condensing boiler, and air conditioning. The total system was close to $40,000. The customer had a much better heating system and he would show his $40,000 heating system to all his friends. It was all because I did not want to prejudge how much money they wanted to spend.

I surrender

It seems like I spend half my day erasing the spam marketing emails in my inbox and it drives me crazy. I had a vendor we purchased an item from enroll me in their email marketing program. I do not mind an occasional email reminding us about their company and believe it is smart marketing. This company's marketing philosophy must have been inspired by a ten-year-old on a car trip asking, *Are we there yet?* It started by sending one email a day and then progressed to two a day. My breaking point was reached when I got three emails in one day. I clicked on the site to unsubscribe. I will never do business with that company again.

Use stories

People are intrigued by stories and these stories may be able to help you to sell yourself and your company. I like to tell clients about how the Spanish Flu affected the design of heating systems and how the Living Room got its name as a result of the flu. Some of these stories are in the rear of the book. I also like to tell people how the industry adopted the standards of 2 Psig for low-pressure steam and 180 degrees for the hydronic systems.

Know your products

Customers want a company who can provide a reliable product. As a representative of your company, you should be the expert. Study the information on the products you sell and be their expert reference point.

What to take on a sales call

Many sales professionals will take a large bag filled with sales aids into the call and look like every clichéd sales person in history. I like to just take a tablet and a pen. It makes me focus on the customer. The other consideration with taking a bag is the customer will wonder what you have in the bag and not concentrate on what you are saying.

Should you include model numbers on the quote?

I am hesitant to include model or part numbers in my sales proposal as it will only take a few minutes for the customer to get on-line and see what the cost is for the unit. They will assume the difference in price between the cost of the unit and your sales price is all profit and they will feel like you are cheating them. If they ask you why it is so much more expensive, I like to remind them the markup includes insurance, call backs, and vehicle costs.

Building Rapport

One of the issues you have when doing an initial sales call is the customer usually does not want you there. Before meeting with a client, I will do some preliminary background checks on the person as well as the company. You could do this with social media. I would be hesitant to bring up personal items as they may feel uncomfortable. I may ask them something about the company. The best way to build rapport is to ask questions about the person. I would ask open ended questions which require more than a single word answer. Some of the questions I may ask are:

How did you end up here?

How long have you been doing this?

Why are you talking to us?

Who did you work for before?

If it is a commercial real estate company, I may ask how their occupancy is doing or how difficult it is to attract tenants in a tight market. If it is a hospital, I would ask how their census is. It is the occupancy term they use to describe what percentage of patient rooms are filled. Be careful your questions do not come across as an interrogation session. They should sound natural and unthreatening.

Nice Fish

I had a meeting with the facilities director for a local real estate management firm and when I sat in his office, I saw a huge stuffed fish on the wall above his desk. I didn't ask him right away about the fish and we talked about his background and his requirements. Finally, my curiosity got the best of me and I said: *"That is one huge fish you have hanging over your head."* The man shrugged his shoulder and I asked *"What?"* He told me his boss is an avid fisherman and he put his trophies in every office. *"Quite frankly, I hate it. I appreciate you not making a big deal about the fish. Every salesperson who walks in here makes a huge deal about the fish and then tells me their own fish story. I hate fishing."* The man said and laughed. Do not assume your contact did the design in his office. I like seeing pictures done by the contact's children hanging on the wall and will jokingly ask, *"Did you do that? It's very good. You stayed inside the lines."* It makes the customer laugh and breaks down the wall. People always want to talk about their children.

My team is better than yours

I knew a sales person who was a die-hard sports fan of the local university and wore his team colors and logo on everything. He walked into a sales call and the client was an equally passionate fan of his school which was the rival of the sales person's school. The appointment did not go well and my friend was asked to leave. He felt it was more important to argue about sports teams than to do business. Leave your school colors at home and try to be neutral. Save your passion for game time.

Cursing

I would not curse when on a sales or service call as it could offend the potential customer. I hired a talented tech who had terrible anger issues. If the service call did not go as well as he thought, he would unleash a tirade of curse words and threw tools across the room. I would get calls from customers informing me of his temper and asking he never return. I tried speaking with him, warning him, and finally let him go.

Sales Presentation

I attended the Dale Carnegie Sales Course when I started on the sales side of the business. The course was very valuable and helped me start to be a sales person. One of the things it did was to show me how to construct a sales presentation. I was very proud of my sales presentation and it was semi-successful. It was built on saving energy as the price of the utilities had spiked. I went to visit a rural school and when I was done with my presentation, the man looked at me and said, *"You did not ask me anything about my needs. I do not want your company to do the service here. For your information, we have a gas well on site and get free gas."* I thanked him for his time and left with my tail between my legs. I was embarrassed and angry at myself but I learned a valuable lesson that day; ask questions. I also learned to have a couple of different sale presentations depending on the answers the customer gave me. I had one for energy savings, one for comfort complaints, and one for lowering repair costs. These covered the majority of my sales calls.

Courage is resistance to fear, mastery of fear, not absence of fear. Mark Twain

Find pains

I attended the Sandler Sales Courses and they had a unique method in their sales training of using questions to find emotional "pains" with the customer. It was their belief people were more likely to act to resolve pains than for benefits. An example the trainer used was a person went to get his yearly physical examination. The man was angry at having to take time from his busy schedule for something he felt was a waste of time. During the exam, the doctor placed the man's chest x-ray on the lighted viewer. The doctor pulled his glasses off and looked at the x-ray like it was the first time he had ever seen one. *"Hmm,"* he said and pointed to a place on the x-ray. *"What's wrong?"* the man asked excitedly. *"Probably nothing. Do you have discomfort on the left side of your chest?"* the doctor asked. Whether the man feels discomfort is irrelevant, he is now scared and nervous. The doctor then says, *"I need to do some more tests."* The patient went from being angry for having to go to a routine exam to a blank check to heal him. The patient felt the "pain" and the sales trainer suggested you need to find at least three of those pains on every sales call. The pains we may find on a sales call could include comfort complaints, breakdowns, or safety concerns. An example may be the feeling the maintenance director has when he or she sees on the telephone caller ID the extension of the chronic complainer.

Questions are the key

If you want to know the key to your customer's needs and wants, ask questions. A way to become a good sales professional is to watch interviewers on television. A sales trainer taught me the key to get a more honest answer is to ask three questions on the same subject. My questions may look like this:

Me: Who are you using for service now?

Client: I am using XYZ (Big Corporation)

Me: Wow, they are a huge company. Are you happy with how they treat you?

Client: Oh it's ok. You know we are a small customer and they take us for granted.

Me: Help me understand what that means.

Client: They never call before they come and they expect us to drop everything to appease them. It really ticks me off.

When asking questions like the ones above, it has to come off as natural and unrehearsed. You also do not want it to be like an interrogation. It takes practice.

So as I consider my sales presentation, I explain in my presentation how my company would never take them for granted and always contact them before arriving. I write myself a note for in their files. During my interview with the customer, I try to find two to three critical sales keys and base my presentation on those.

What are some common questions to ask a residential customer? The following are some of the questions I ask when interviewing a homeowner.

Is there much of a temperature difference between the top of the house and the first floor?

If there is, you could consider zoning, air balancing, or continuous fan operation.

What room is the hardest to heat and or cool?

This assumes they have an issue and you could provide a solution.

Is the furnace noisy when it starts?

When I was a child, our furnace was so loud we would have to turn up the television volume just to hear the show. Once the furnace shut off, we would have to lower the volume. This was before television remote controls so the job of adjusting the volume up or down fell to my siblings or me. My proposal would have included a variable speed fan and some duct modification to quiet the unit.

Why are you looking to replace the unit?

I was asked to provide a replacement cost for a new "furnace" for a recently widowed woman whose son played soccer with mine. When I arrived, I asked the woman why she was looking to replace the furnace. She said the company who did her clean and check told her, *"The furnace is unsafe. The heat exchanger is rusty and cracked and may be throwing dangerous carbon monoxide into the house."* I went into the basement and was surprised to see a hydronic boiler and not a furnace. I checked the boiler and it worked great. The unit was only ten years old which is relatively new for boilers. I checked the flue gases and found no dangerous carbon monoxide in the flue or basement. I went upstairs and explained this to the woman and her neighbor who asked me to look at the unit. She thanked me profusely. I wanted to call the owner of the company and let him know what I found and explain how sleazy his tech was. I decided against it and was glad this tech worked for my competition. As a result of this service call, we got the service at several of her neighbor's houses and the woman told anyone how well we treated her.

This question can help you to customize your presentation to what is of interest to the customer. One customer told me the reason he was looking to replace his furnace and the air conditioner was because the existing one broke down "all the time." With further questioning, I found out when the unit broke down, he had to send his employees home and lost money that day.

During my sales presentation, I stressed the furnace had an extended warranty and if it did break down, our service department is manned 24 hours a day. We would be there in a well-stocked truck and get him back on line quickly.

How reliable is the furnace and or air conditioner?

This will help you discuss a new unit with an extended warranty

Does anyone have allergies or asthma in the house?

You may be able to include a humidifier, air cleaner, or UV light in your proposal.

Do you ever wake with a stuffed nose in the winter?

A humidifier could solve this problem.

It's your problem now

Once you install a new unit, all the issues the client had are now yours. I have been blamed for everything from the light in the garage stopped working after the furnace was installed to the pet hamster escaping when we installed the furnace.

How to Listen

Listening is a fine art and takes practice. You should make eye contact without looking like you are staring the person down. You also want to provide feedback to the person so they know you are paying attention. This may include nodding, smiling, paraphrasing what they just said, writing a note on your pad, or asking a question to clarify what the person was saying. Some more ways to elicit better responses is to say Really, That's interesting, or Fascinating.

Want to learn how to become a great interviewer?

If you want to become a skilled interviewer, try to hone your skills by interviewing a teenager. They are the most challenging humans to talk with and if you can hold a conversation with a teenager, you will do great in sales.

Help me understand

A positive way to help you understand what they are looking for is to say, Help me to understand... This could be the point they just made or it could be something they just said. You could repeat it back to the customer as well. So as I understand it, you are looking for...

Some clients will try to minimize what they said when you ask a question about it. They may say Oh Never Mind. If you ignore it, do so at your own risk. It is usually something significant to the customer. I would lean forward and say, *"No, please help me understand. I can sometimes be a little dense."* This will endear you to the customer.

Sympathy vs. empathy

As a sales professional, you have to understand the difference between empathy and sympathy. We want to show empathy and not sympathy to the customer unless they are a good friend or relative. To exemplify this, consider going to a doctor and she tells you have a serious health issue. The empathetic doctor will comfort you and tell you she will be there the entire time. The sympathetic doctor will cry and wail, *"Oh what are we going to do."* You want the empathetic doctor and not the sympathetic one. The customer wants the empathetic sales person.

Stay out of my space

I always joke with my kids to stay out of my space unless I invite you in as it makes me feel awkward. The customer has a personal space and I think you have to respect it. A sales person I know will take the chair and move it from in front of the desk to the side of it when he is on a sales call. He feels it breaks down the barrier between the customer and himself. He tells the customer he has some hearing issues and this helps him hear. I tried it several times and never had the success my friend had.

Edward T. Hall, an American anthropologist, developed the concept of "Proxemics," a study of personal space. His theory is based on people are allowed into personal spaces according to how well you know the person. It is comprised of four zones. The first is Public Distance Zone, and it is usually greater than twelve feet away. This is for public speaking and addressing a group. The second zone is Social Distance Zone, and this is between 5-10 feet away. It is where you would start a conversation with someone you do not know very well. This is the zone most people prefer when first speaking with a sales person. The Personal Distance Zone is between 2-5 feet away and is reserved for friends and family. The more you like someone, the closer they stand to you. The last zone is the Intimate Distance Zone, and it is less than two feet away. This is for the most loved and trusted, usually siblings and partners. If you, as a salesperson, go inside this zone without the customer's consent, the customer's mind and body will react negatively.

Monkey's paw

Sandler Sales uses this analogy in their sales training and I liked it. When a large ship is docked, it is secured to the dock using thick ropes capable of holding the ship in place. These ropes are very cumbersome and hard to manage. To be able to maneuver the rope from the boat to the dock, the sailors use what is called a monkey's paw. The monkey's paw is a smaller rope with a knot the size of a fist on one end. The other end of the smaller rope is attached to the larger rope. The crew member on the ship will throw the monkey's paw knot to the dock hands. The dock hands will pull the small rope and the large one will follow. This analogy works well with sales. The idea is get some small commitment from the customer to prove your value to the customer. For example, I may ask a real estate management firm to allow me to look at one of their systems and see if we can get it working. If we can, hopefully they will use us for their other systems. This arrangement works for both the customer as well as our company. What if we find the customer is very difficult to work with? Or does not pay their invoices promptly? If we had a large contract for a year or longer, it might be a nightmare. This allows both of us to test the water to see how we work together. An example of this is when I offered to look at the boiler for a flat fee of $99.00. I explained it was a one-time offer and they agreed. I found the issue was the pilot assembly. I cleaned the pilot assembly and the flame failures stopped. We got along well with him and his company and they gave us the contract to service his boilers.

Last one there

When you are inside a boiler room, take a few minutes to walk through the boiler room and look around. In one boiler room, I looked around and saw the flue for the water heater was peppered with holes and rust. I shut down the unit and called the building owner. He stopped by and thanked me for finding the issue. I did it as a service to the client but I also did it to protect my company. If something happened, they could say I should have caught it. They always blame the last one there.

Reversing

This is a sales technique where you use the client's objection as part of the sales process. For instance, if a customer says *"Your company is too small."*

To reverse it, you say *"That is exactly why you should go with us. We will be more responsive to your needs and you will always be able to talk with me."*

Another example is if the person says *"You are more expensive than the current service provider."* I would answer, *"That is exactly why you should consider us. We have highly paid and highly trained service technicians so we can make sure your system works properly and safely."*

Know your competition

Be aware of your competition and how they go to market. For example, there is a local company which features these quirky commercials on the television and radio. If I am competing against them, I will mention the ads with something like *"How much do you think they spend on those ads? We looked into advertising and it is very expensive. We decided to save the advertisement costs and pass the savings to my customers."* Use their own sales tools against them. For example, another big company which sells many units competes against me. I will use their size against them. I will mention they are really large and have a huge payment each month with all their people and trucks. I suggest the customer would be just another number to them but to me, they would be a valued customer as I personally oversee each installation.

Trouble shooting sheet

A contractor I know makes up basic troubleshooting sheets printed on fluorescent paper and laminated and attaches them to the equipment. Each sheet will contain basic things to check and the company phone number to call in case something is wrong. It is a smart marketing idea to keep his company name in front of the customers.

Don't give away your expertise!

It is rumored Pablo Picasso, the famous artist, was sketching in the park when a woman approached him. She asked him to draw her portrait. Picasso agreed and sketched her portrait with a single pencil stroke. *"It's perfect. How much do I owe you?"* The woman asked.

"Five thousand dollars." He answered.

She was shocked and asked why it was so expensive explaining it only took a few seconds. Pablo Picasso answered, *"Madame, it took me my entire life."*

Think about all the dirty nasty jobs you have seen in your career. They have been learning experiences for you. You are the sum total of all those jobs. Many techs will give away their expertise. I hope you do not do that. Always charge for your expertise

You are a professional and should be paid for your knowledge. Some clients will invite you to their facility and hope you can diagnose a problem for free. While on site, one client asked me to look at his boiler with intermittent problems and wanted my opinion. I told him we could schedule a technician but he said: *"I already spent lots of money on this boiler and did not want to spend any more."* He wanted me to provide a free service call and help my competition at the same time. That takes guts I thought. I wanted to ask the man to give me $100 from his wallet and when he would balk I would say that is what you are asking of me. A warning if you are a sales person and do service, the insurance company may not cover you. The insurance costs for a service technician is much higher than the insurance costs for a sales person.

You can close more business in two months by becoming interested in other people than you can in two years by trying to get people interested in you. Dale Carnegie

Look at this will you?

A commercial HVAC business owner told me a story about one of his younger techs was doing free work for his family. The tech asked the business owner to look at a drawing he made to change the wiring for the heating system. The business owner quickly looked at the drawing and agreed it would work. During the winter, the unit failed and the relative called the business owner and threatened to sue because he reviewed the changes. The owner had to repair the system at his cost. Be careful doing pro bono work.

Two ears and one mouth

My grandmother Madeline used to say *"You have two ears and one mouth. You need to listen twice as much as you talk."* This advice works perfectly for a sales call. Ask questions and listen to what the customer says. The sales presentation is all about the customer not you. I do not wish to be mean but most clients do not care about you. It is nothing personal. It is just they are tied up in their own lives. Your sales will grow if you are more interested in the client than yourself.

Chinese Restaurant and Side Work

I went to a Chinese restaurant for some takeout and was wearing one of my company shirts with our logo. The reception area seemed to be a bit warm. While waiting for the order, the owner of the restaurant asked if I could do some side work for her. I suggested she could call my office and we could schedule a tech. She shook her head and said it would be too expensive. It angered me and I asked if she could give me the name and number of her chef. When she asked why I told her she was too expensive and I was hoping her chef could come to my house and cook dinner for my family. She had a shocked look on her face and walked away to get my order.

Can you top this?

One of the games we humans play is, Can you top this? It involves the customer telling you something about themselves and you topping the story with one of your own. I urge you not to do it as it will back fire. It will minimize how the client feels and make them feel less worthy. If you have a similar experience as the client, use the opportunity to ask questions about how they felt or dealt with the experience. For example, the customer tells you he was having issues with his vehicle. Rather than talking about your own vehicle woes, ask questions to show you are empathetic to their predicament. The other key to this is not to offer your solution for their problem. Again, it will make them feel less worthy and could anger them. I would ask them what their solution to the issue is.

Use facts and figures

When using facts and figures in your presentation, use legitimate sources such as ASHRAE, recognized industry experts, or government sources. I would refrain from quoting sites which may be controversial.

It's not about you

Some sales people feel compelled to push their opinion on the customer. I was out with a factory sales person on a sales call. We were talking with the client and she mentioned something about Global Warming and the carbon footprint of my boiler. This set the salesperson off and he felt it necessary to inform her that her beliefs were based on fake science and she should get better informed. I cringed and the woman ended the sales call abruptly and we did not get the sale. I was furious and told the sales person he should keep his opinions to himself.

Rumors

If you have been in the industry for any length of time, you have heard rumors about a competitor. There are most likely rumors going on about your company. I have one particular competitor who tells everyone I am going out of business. It's been going on for 30 years. I will simply laugh and tell the customer, *"They wish I were going out of business. I have taken some of their business and they are mad."* Do not fall into the trap of spreading rumors. That is not the way you want to grow your business. It will backfire on you and you are better than that. You do not want to be that person.

Forget what I told you about the last company

One of the reasons people will buy from you is because they trust you. If you go from company to company, you lose some of your credibility. You initially sold them on the first company and now you walk in and tell them, *Forget what I told you before, this is really the best company.* If you leave there and go to another firm, you lose even more of your credibility. I have heard the magic number is about three employer changes. If you do more than that, customers will hesitate to switch with you.

When you do change to another company, do not slander the old company. It will make you look bad and I have seen sales people go back to the old company. If you do change employers, I would suggest you say something positive about the old company. I would say something like, *"It was the most difficult decision because the (Old Company) treated me we well. This (New Company) offers me better solutions I can share with my clients. I wish my old company well and appreciate all they did for me."* Be positive about your previous employer even if leaving was not your choice.

Get 100 Facts about you and your company

A homework assignment is to try to come up with as many facts as possible about you and your firm. If you are looking for facts about yourself, it may be anything from how well you did in school, to how much training you have, awards you won, to even how your last company liked your service. Now do the same thing for your company. This may include how long you have been in business, how much insurance you have, or how many customers you have. I would write these in a spiral bound notebook you can keep with you. You will not get this done in an hour or even a day. It will be a work in progress. Be sure to skip a few lines in between each fact. I urge you to hand-write these facts because handwriting allows you to remember the facts better than simply typing.

Facts bore benefits sell

Now comes the hard part, you need to assign a benefit to every fact you have written. There is a famous adage by the famous author and sales trainer, Dale Carnegie who observed millions of ¼" drill bits are sold each year. He surmised the purchasers bought the bits not because they wanted the bits but rather wanted the ¼" holes. The drill bit is the fact and the hole is the benefit. After each of your facts, you need to ask yourself, What does that mean to my customer? The reason you need 100 facts and benefits is to be able to respond to any issues discovered during the questioning phase of the sales call. Consider this example:

During the questions, the owner talked about the reliability of his air conditioner. He told me the story of how his ac unit failed on July 4th and all his relatives were stuck in the house and had to gather fans to try keeping them somewhat comfortable.

Fact: Our preseason air conditioning check covers 79 points and every major component.

Bridge: This means

Benefit: Our thorough testing will uncover any weak or questionable components and the odds of a failure are greatly reduced.

A bridge is a transition between facts and benefits. It is usually something like so, which means, therefore.

How many fact and benefits do you need?

I like to use at least 3-4 facts and benefits for each sales call but use as many as you need to sell the customer. Be sure they pertain to what interests the customer. You will know these from the questioning period of the sales call.

All progress takes place outside the comfort zone. Michael John Bobak

Could insurance be a benefit?

I was meeting with my insurance agent for our yearly review and he told me a story about a homeowner who hired a roofer "under the table" to repair his roof. The person he hired was a maintenance technician for a local school and he had no worker's compensation insurance. During the repair, the roofer slipped and fell to the ground and was severely injured, unable to work. He sued the homeowner and the homeowner's insurance had to pay the man's wages. Shortly after, the insurance company canceled the policy and the homeowner's rates were substantially higher with the next company. I used to tell this story when I talked with my customers about our insurance coverages. Some people like to hire a person under the table to do their work. I would show our insurance policy and assure the customer they had nothing to fear like that with our insurance.

Never slam the competition

I never like to slam the competition because they have said some pretty nasty things about my company and me. If someone asks me about a competitor company I disrespect, I would say *"They have been around a while."* Or *"They are a big organization."* If the customer would like to slam the competition, I will not stop him or her and may even subtly promote it. After all, I am doing the customer a favor by allowing them to vent. Something as simple as asking *"Really?"* could allow the customer to open up.

Ask open ended questions

Try to avoid asking closed end questions which can be answered with a single word. Instead, ask open ended questions which will elicit a longer response. Instead of asking, *"Do you like your existing service company?"* which will result in a yes or no response, try asking something like this:

"What do you like about the way they service your company?" This will help you to show what you can do for the customer in your sales presentation. You can see what interests them.

Now comes the inverse where you can get substantial sales ammunition. Instead of asking *"What is it you do not like about what they are doing?"*, I would ask *"What would you change about how they service your building?"* This is a little less combative and allows the customer to open up. If you listen closely, they will tell you the keys which are important to them.

The following are some good open ended questions which can allow you to uncover some issues which you can solve.

What areas are the most difficult to heat or cool?

If you could change one thing about your system, what would it be?

If you had a magic wand

This is one of my favorite sales questions I use. I realize it sounds a bit hokey, but it works. For example, you may ask If you had a magic wand, what would you change about... your system? your service? If you say it with sincerity, it will allow you to know one of the keys you will need to sell the customer. Try it and you will be pleased with the results.

What would the perfect system look like to you?

This question which appears hokey does work. It allows the customer to describe what is important to him or her.

What have you tried so far?

Once you find the problems or issues the customer is experiencing, I like to ask, *"What have you tried so far?"* and listen to them. Once they explain what they have done to stop the problem, I will ask *"How has that worked out?"* This allows them to evaluate their efforts and you to work on a solution.

No one is perfect

I like to use this when the customer is telling me about how wonderful the current company is. I will say, *"They sound almost perfect and I know no one is perfect. Is there not one thing you would change about them?"* I will say this with a smile and wait for a reply.

What do you think would resolve the problem?

This is another question which allows customer input into the resolution of the problem. You might be able to incorporate their solution into your presentation. If you do, I would mention we are using one of their solutions.

May I take notes?

I believe this is a powerful sales tool and will take a tablet in with me. I will always ask, *"May I take notes?"* Most people will smile and agree. It makes them feel important and they are important.

Summarize your questions

When done with your questions, I suggest you summarize and get the three most important items. For example, I would say *"In summary, the items you are hoping to resolve is 1,2,3. Is that correct? Wait for response. Then ask "Is there anything else I missed?"* Occasionally, a customer will say something I did not think was important but the customer did. These keys will form the foundation of your sale presentation. This may sound something like this:

In summary John, the three things you would like resolved are:

Even the heating out between here and the master bedroom

You would like a more quiet system

And we need to do something about the dry air every morning

Is this correct?

Client: *Yes*

Is there anything else I missed?

Client*: Yes, I want to be able to monitor our system from my phone because we travel often and I want to make sure my house does not freeze.*

"Thanks, I almost forgot that" I would respond. If the spouse or partner is there, I would ask, *"Is there anything else we forgot?"* This allows the spouse or partner to respond with what is important to him or her.

Now I can do my sales presentation and use these as call backs. Call backs are what comedians use to drive home a point or joke. I would say *"We are suggesting a variable speed blower which will circulate air all the time and will eliminate the temperature difference between here and the bedroom. You were saying this was a concern for you."* I covered the customer's concern and will do it with all the important issues. At the end of my presentation, I will again summarize how my solutions will solve his or her problems. It may sound like this. *"Mrs. Smith, we have talked about the issues with your current system and how we can make the temperature difference go away, the new system will be quieter, the morning nose bleeds will be gone, and you will be notified on your smart phone if the temperature is not within your parameters. Do you see how this will work for you?"*

What if the customer comes up with something new? I would ask questions about it and see how important it is to the customer. For instance, I may say something like this: *Ms. Client, I must have missed that during our talk. Please help me understand what you are looking for here.* This throws it back to the customer and allows them to tell you another sales point they feel is important.

There's just one more thing

There was a television show called Columbo and you can find it on-line or on Netflix. It features a disheveled looking detective who appeared completely incompetent. The criminals would dismiss him as inept and when the show was about to end and the criminal thought he or she would get away with the crime, the hero would solve the case by asking one more question. I used to style my sales questioning technique after his. If you want to see an expert at asking questions, I urge you to watch a few episodes.

One call or two?

When you are going on a sales call, you have to know whether it is a single sales call or longer. For example, if I am going on a commercial sales call, it may be two or more sales calls. The first may involve talking with the customer and then do an inventory of their equipment. The second sales call may be the sales presentation. On a residential sales call, I will limit it to a single call as the sales volume is not as large.

Two call ideas

When making commercial sales, it usually is two or more sales calls. The first sales call is to ask questions and assess the customer's needs. The second is when you present your solution and sales presentation. When you start the second or final sales call, I urge you to repeat the issues or "pains" the client said in the first meeting. After the small talk, I would say something like this, *"During our first meeting, you said the following issues were important."* I would then list the three most important points we discussed in the first sales call. Ask if those were correct and if they have any other important issue I should cover. After that, you should start your sales presentation.

It's ok to say no

Many sales people are afraid to hear the word no. When making a sales call on a one visit sales call, I like to tell the client in the beginning, *"The only thing I ask is you give me a yes or no decision today and it is ok to say no. Is that good?"* You have to do this with compassion and not arrogance. When the sales call is done, I will ask the client, *"Does this meet your needs?"* Customers are afraid to tell you no. They may say something like *"I need to discuss this with my spouse or significant other."* I may ask them *"Does this mean no? It's ok to say to no."* The idea is to find the real reason they are not going with your firm. Have fun with this. If the client says no, I will close my folder and thank him or her for their time and then say *"Just curious, what was it which made you decide to choose the other company?"* The mere act of closing the folder will cause the wall to come down and disarm the customer in most instances. This may allow you to find the real reason why the client did not choose you.

Questions

When I first started in sales, I used to have questions listed on a sheet of paper. I learned to limit the number of questions on the sheet to 4-5. If you have more than that, it is intimidating to the customer. I had a client ask, *"Are you going to ask me all those questions? I don't have time for that."* As I gained more confidence, I memorized the questions and they sounded more spontaneous.

List of customers

Should you keep a list of customers as a sales tool? I believe the list should be protected like gold because if your competition gets your list, they may be able to target and steal some of your customers. I keep a list inside an IPad or notebook and it is never left with the customer. If the client asks for a copy, I would tell them I would rather keep the security of our customers safe and could not do it without asking every one of the customers.

Sales Truth

One of the things I found after being in the trade for over three decades is this: A client will act quicker for a comfort complaint or a breakdown than they would for energy savings. While most clients will tell you they are interested in being green, it turns out most are more interested in stopping the comfort complaints or the breakdowns. It is after you assure them they will have a reliable and comfortable system that they will consider the environment.

I have no special talent. I am only passionately curious.
Albert Einstein

Body Language

This is a strange science, and many people are sold on the legitimacy of it. Sophisticated buyers will arrange to have their desk higher than the seat on the other side of their desk in a way of exerting control over the person. It is believed a closed physique such as arms folded across their chest are showing closed minded thinking and not open to you. One of the ways to deal with people like that is to hand them a sheet of paper they have to hold. I know I said never hand the customer anything but you need to open their stance so they are willing to listen to your presentation. I like to place the paper on the desk so they have to lean forward and open their arms to read the paper.

How long should you wait?

One of the things which drives me crazy is waiting for a customer or prospect when we have a set appointment. I understand things occasionally happen that are out of control but sometimes people are just plain rude. One of the big factors here is the size of the potential business. Some buyers will make sales people wait to see how serious they are about getting their business. I usually limit my wait to fifteen minutes. As it gets close to ten minutes, I explain I have another appointment and will have to leave soon.

Is it a test?

I read where a person would text or call a job applicant at night or on the weekend to see how quickly they respond. If they respond promptly, they are considered for the position. A buyer once told me he calls a potential vendor in the middle of the night to see how well they respond to an emergency situation. If they do not respond quick enough, he will not consider them.

Be on time

The famous football coach Vince Lombardi had a saying, If you are five minutes early, you are already ten minutes late. He believed in being 15 minutes early for everything. It is often referred to as Lombardi Time. When you have an appointment with a client, always try to be 15 minutes early. It will give you a chance to collect your thoughts and prepare for the sales call. It is rude to make your appointment wait. Being late will adversely affect your sales.

What to do if the client is on the phone the whole time.

I have only been on a few sales calls where the owner kept taking phone calls and it frustrated me. I would be in the middle of a question or presentation and the phone would ring and the customer would answer it. I was afraid of calling the person on it because it now becomes adversarial. What I may do is ask if there is somewhere we can grab a coffee or lunch. I figure if I can get the person away from his or her phone, my sales talk will go much better. I had an appointment with a client who wore a nearly invisible ear bud. You would be in the middle of a conversation and the client would start on an entirely different track. I thought he was possessed and then I saw the ear bud. He apologized but it knocked me off my game.

Keep the technical talk limited

I know you are smart but try to limit the technical talk. Many people will not understand it and you will lose their concentration. Try to come up with ways to make what you are saying understandable.

Using Props and Showmanship

Props can be an effective sales tool if used appropriately. An electronic tablet may be able to help you develop some demonstrations to illustrate your points. Just remember these props will not sell your service or product, that is up to you. I was one of the first sales people to talk about modular boilers in my area and many of the engineers found it a difficult concept to accept until I constructed a demonstration panel consisting of 4 incandescent light bulbs mounted on a piece of wood. I lit all four light bulbs to show the heat required on the coldest day and explained how less heat is required as it warms outside. I simulated it by switching off one light at a time. After seeing this, they got it. It was crude but effective. I was known as the boiler guy with the lights.

Showmanship is almost like bragging and be careful using those in a sales presentation. An example of showmanship is when you show a letter of recommendation given to you by a past customer. It could also be an award your firm won.

Talking about your company

Your sales process should be based on the answers the prospect gave you during the interview process. At some point in the sales process, you will have to talk about your company. For every fact you give them about your company, you need to give them a benefit they can relate to. A question I ask is *What does this mean to my client?* after each fact. For example, *We have a combined yearly experience of 60 years. This means we have seen most problems and should be able to fix anything we see in your buildings*. Another example might be, *Our phones roll right to my cell, so I will be able to answer your call in case of an emergency.*

Change your thoughts and you change the world. Norman Vincent Peale

Word Picture

A word picture is the emotional hook you use to sell you and your product. It is a short story where you illustrate something about your firm and I like incorporating them in my sales presentations. You can either use either a positive or negative word picture. A positive one may be about someone happy when using your company and negative one may be one where the person did not use your company. The following is an example of a positive word picture.

The Acme Realty Company had us look at one of their problem buildings. It felt cold and clammy when we walked in and smelt like mildew. The temperature difference was so bad there was a 15-degree difference from one side of the building to the other. On one side of the office, the ceiling diffuser was blowing the papers off the desk and cold air was blowing onto the necks of the people working there. The people on the other side could barely feel any air, feeling stuffy and warm. The staff called the boss 4-5 times a day. We balanced the airflow and they were amazed at the temperature difference from one side to the other was less than 3 degrees and they found their utility costs dropped by about 5%. Also, the comfort complaints were slashed. See how I got many different senses into the word picture; smell, feel, sight.

I was teaching a class on word pictures and one of the attendees was a life insurance sales person. He chose to use a negative example in his practice sales talk and it left me speechless. If you knew me, that never happens. His word picture was geared to a husband and he said, *"If you died without enough insurance, your wife and family could be homeless living on the street. Is that what you want to happen to them?"* Now you can see why I was speechless. Negative word pictures sometimes are effective but in this case, it was over the line. I suggested he use a positive word picture.

An example of a negative word picture in our industry might sound like this: *A school district just north of here discovered a small leak on their only boiler in the elementary school in late summer. They did not have the money to replace the boiler and hoped to just get through the heating season. Sure enough, the leak worsened and the boiler failed on the coldest day of the year. Don't they always? They were forced to cancel classes and get an emergency boiler in place so the pipes did not freeze. The costs were 25% higher than if they had done it in the summer. The Maintenance Director was really on the hot seat with the board. Now, you would not want to go through anything like that. I would suggest you consider having your boiler checked before you need it.* This was negative but it was pointed toward someone else.

Under promise over deliver

When talking with your customer, I like to under promise and over deliver. What does that mean? If I tell the customer they will see a 30% reduction in their utilities and they only see a 25% reduction, I look like I lied to them. If I underestimate the savings and promise a 20% reduction and they see a 25% reduction, I look like the hero and their advocate. They still saved the same but we have two completely different feelings.

How much is in the budget for this?

After building rapport with the client, I like to ask this question. It allows you to see how much the customer is looking to spend. Some clients will answer it honestly and others will not. In some instances, they will say something like, As little as possible. After their answer, I would go into the bracketing technique.

Bracketing

I like to use price bracketing to see how much money the customer is thinking about spending and it also eliminates the pricing objections later. Let us assume the installed price ranges from $3,000 for an on off furnace to $5,000 for a variable speed high-end model. For instance, if a customer was looking at a new furnace, I may ask something like this: *"From what you told me, you are looking for a reliable heating system which will lower your heating costs and increase comfort. A new furnace will run anywhere from a couple thousand for a standard on off unit up to $6,000 for* *a top of the line comfort system. So I can better provide you what you want, where are you in this range?"* Now, all the customer heard was $6,000.00 and that is what they will remember. If your high-end furnace comes in below $6,000.00, you are a hero. Usually, the customer will say something like I do not want the cheapest but I don't need the best either. So we know we are in the same area. If the client says, I only want to spend $1,500; you have a problem. I would address it right away and tell the person we will not be able to meet that price. I would ask something like, *"Is it over?"* This will tell you how adamant they are about the price. I would also ask *"Have you received a price that low already?"*

"Is it over?" is a sales tool question I like to use when speaking with a customer. If you are far apart on a point or price, it allows you to see just how important the point is with the customer.

No Buts

Be careful using the word But in your sales presentation. The word But negates everything you just said before saying the word. To exemplify this, I had a customer with a defective gas valve on his five-year-old snow melt boiler. The person does no maintenance on the boiler. He called and was upset and said "I ripped him off."

I could have two responses and you can see which one

works better.

Response 1

"Mr. Customer. I understand your frustration. I would be angry if a five-year-old piece of equipment did not work. What would you like me to do?" The customer would most likely say, *"Please help me get it running. We have a snow storm forecast tomorrow."* The customer is ok and feels like I heard his concern and worked to solve the problem. Now, look at a different response:

Response 2

"Mr. Customer. I understand your frustration but you never do preventive maintenance." Now, instead of me being compassionate, the customer is angry because I insulted him, even if it was true.

I think Response 1 is a better response.

Is it based solely on price?

I have a theory; an uneducated customer will always make their choice primarily on price. Your job is to educate the customer about you or your company. Very rarely is a decision based solely on price.

Just give me a ballpark number

I hate when clients ask for a ball park number on the phone as it will always come back to bite me. I use humor to disarm the client. Since I live in Pittsburgh, PA., I will answer it this way: Well, Heinz Field cost $281 million and PNC Park cost $216 million but your system should be much lower than that. For me to give you a proper estimate, I will have to look at the old unit and perform a heat loss on the house to be sure we have the right size unit.

Why am I here?

Somewhere in history, the cavemen decided we need to get 3 or more prices when purchasing an item. Are you the check price they need so they can award the job to their current service company? Perhaps I become jaded after being in the industry for this long but I like to ask the client why they asked us here. They will probably say we are there to give them a price on the equipment. I will ask the client who they have talked with already. When they tell me who they spoke with, I ask what have they told you. I will do this gently of course. Another question I would ask is:

"They are a good company. Why did you not choose them?"
This helps me understand their needs and or wants.

When are you looking to have this done?

This is an important question and it allows you to know whether this is something they want now or if it is just a project they are looking at for the future. If they are looking for something in the future, I may say, *"Prices may change between now and then. A budget price will be between $3,000 and $6,000 depending on options you choose."*

If they tell you they are looking for this right away, I would say, *"I have to check our schedule but I may be able to switch some things around."* Now this tells the customer you are busy and willing to work for their business. I tell them this even if we are slow. Everyone wants to work with someone busy. It also gets an IOU from the customer because you are willing to switch things around for them.

How does the decision process work?

When going on a sales call, you should find out how the decision will be made. If the person tells you another person makes the decision, you may want to ask if you could arrange a meeting with that person. If you are in a home, the person may tell you their spouse or significant other makes all decisions. They may say we make them together. In any event, your chances of a sale significantly drops if you are not in front of the decision maker. I will ask if we can call the other person and get their feedback and make sure we have all their concerns covered.

Using referrals

Using referrals or famous names can be a bit tricky. For example, if the prospect dislikes or does not respect the person you are using as a referral, it could backfire on you. If they like or respect the person, it could help your sales process.

The following is an example of using a referral where the person is respected:

John Smith with American Real Estate called us about a building with no heat. Do you know John Smith?

Client: Yes

Do you respect his opinion?

Client: Yes

He said we are one of the top HVAC companies in the area.

The following is an example of using a referral where the person is not respected:

John Smith with American Real Estate called us about a building with no heat. Do you know John Smith?

Client: Yes

Do you respect his opinion?

Client: No

American Real Estate said we are one of the top HVAC

companies in the area. We hope you feel the same about us in the future. This allows me to move away from the person and toward the company.

The following is an example of using a referral where the person is not known:

John Smith with American Real Estate called us about a building with no heat. Do you know John Smith?

Client: No

He has a position similar to yours and said we are one of the top HVAC companies in the area. We hope you feel the same about us in the future.

Small talk

There is an unspoken barrier between sales people and prospects. When you are first meeting a prospect, you have to engage in small talk to break the barrier. It may be something as simple as asking how long they have been at their position to how long they owned their house. The best way to practice small talk is to talk to anyone who comes within 3 feet of you.

Good, Better, Best

When doing residential sales and service, I like offering three options such as good, better, best. For example, you are giving a price for replacing a furnace. The good option would be a builders model furnace with single speed fan and single stage heat. The better option may be a more efficient furnace with two stage heat. The best option would be the top model for the furnace you sell. It may be a variable speed with multiple stages. When I present my sales proposal, I start with the Best and work my way to the Good. I also offer additional options such as a humidifier, programmable thermostat, UV lamp, high-efficiency filtration.

Help me build your comfort system

If you visit the website for almost any vehicle manufacturer, they have an option for you to build your own vehicle. People love being able to help, especially if they are helping you to help them. I like to enlist the customer's help to design the perfect system for their home. I have a sheet with all the options and slide it to the customer. I let them choose which options they prefer. It works well and they are selling themselves and you are there to guide them.

Getting to know the customer

When I was new in sales, I purchased a book entitled, Swim with the Sharks without Being Eaten by Harvey MacKay. I enjoyed the book and found the best part of the book was the questionnaire he used for each client. It included a 66 question profile about the customer. I made my own and answered the questions after each sales call. It made me more cognizant of the customer and a much better sales professional. Some of the questions included background, spouse and family's names, interests, hobbies, passions.

Find your rainmaker

A rainmaker is an influential well-respected person in the industry who is a fan of you or your company. This person is looked to by his or her peers as a role model and they can influence others to use your firm. They can help you grow your sales volume.

Don't spill your candy in the lobby

A sales trainer I know used to say this line and I never understood the meaning but was embarrassed to ask. I found the meaning when I was talking with a potential customer. The customer asked me about a problem they were having with the controls on an old temperature control system. I looked at it and figured out how to resolve the problem. I tried to explain how I could repair it. The prospect asked me to make a drawing so he could understand my solution. I made the drawing and the prospect told me he needed to speak with his boss and will get back to me. He thanked me for offering a solution. A few days later, my boss called me into his office. He showed me a proposal request and then I saw my drawing was included. It asked me to bid on my solution. I submitted a bid and called a few days later and found out we did not get the job as my competitor underbid us using my solution. I was furious. I met with my boss and he smiled and said *"You got schooled. Don't spill your candy in the lobby."* In other words, do not offer solutions unless you get paid by the customer.

Sometimes clients lie!

This was one of the most painful lessons I learned while in sales. I guess in my naïve world; I thought everyone told the truth. Prospects will tell you your proposal looks "good" to get you to leave. What I do when that happens is to ask the customer, "What does that mean?" This helps to clarify what the customer meant.

Who works there

Always treat every employee in a facility with respect because you never know who is related to who. I walked inside a building and saw a young man struggling to move a cart loaded with parts inside a door. I helped the man and asked where to find the person I was meeting. He showed me where to go and thanked me for the help. I met with the prospect and the young man walked into the office and asked a question. It turns out the young man was the son of the company president and heir to the job.

I learned the maintenance worker or assistant you meet today could be in charge in the future. An assistant I worked well with in a company went to work for another company and I was the first person he called when he needed a boiler.

Using expert opinions in your sales call

I like using expert testimonials on my sales calls. It could be something from a government publication to a chart from ASHRAE to illustrate my point. One of my favorite charts I use is the ASHRAE Equipment Life Expectancy Chart. ASHRAE is the organization who writes the codes and standards for heating, cooling, and ventilation.

The median life is the average life expectancy of the equipment. I will use this when discussing the estimated life of the equipment. For example, you go on a service call and the furnace is 17 years old. Would it be worthwhile to put much money into a system that old? ASHRAE median life for a furnace is 18 years.

ASHRAE also publishes a list of estimated maintenance costs for various heating and cooling systems. It is also a great resource. The following page shows the life expectancy of components in our industry.

Equipment	Median Years
Air conditioner, window	10
Residential air conditioner	15
Commercial through the wall	15
Water cooled package air conditioner	15
Residential heat pump	15
Commercial Heat pump	19
Rooftop air conditioner	15
Steel water tube hydronic boiler	24
Steel water tube steam boiler	30
Steel fire tube hydronic /steam boiler	35
Cast iron hydronic boiler	35
Cast iron steam boiler	30
Electric boiler	15
Furnace	18

What are you selling?

We believe we are selling a furnace, air conditioner, or boiler. Instead, you are selling the following:

- Money (Energy savings, lower repair costs, increased building value, more money to bottom line.)
- Safety of the person, the family, the pets
- Comfort

An old movie from the 1960's, Cool Hand Luke, had a famous line which said: *"What we have here is a failure to communicate."* It is what sometimes happens when a technical person is selling a technical piece of equipment to someone who is not technical. While I would droll if I looked at a state of the art unit with variable speed and a digital readout, most clients would not. They do not care how it was made or what it did. All they are concerned with is whether you and your product can solve their problem. It is your job to find out what the problem is and explain how your product can help.

Most people do not want to purchase a new piece of equipment. They would rather spend the money on anything other than a new furnace or air conditioner. What you are selling is a comfortable home or building in the hot or cold weather, a safe home which will not poison their family.

Is it really an investment?

Some homeowners will laugh when I tell them a new HVAC system is an investment. Let me compare the purchase of a new car to a new HVAC system. Consider a vehicle with a cost of $30,000. When you drive the car off the lot, the value of the vehicle dropped 11% making it worth $26,700. A year from now, the vehicle will have lost 25% of its value and now worth $22,500. In three years, the vehicle will be worth $16,200. After five years, the vehicle will have lost 63% of its value with a worth of $11,100. The vehicle will have lost about $18,100 over five years or an average of $315 per month.

Compare that with a new state of the art HVAC system valued at $15,000. Our new system will save us about $500 per year in utilities. After five years, our furnace is only 1/3 of the way through its estimated life and we have saved $2,500 or about $41.67 per month in utilities and increased the comfort. I think saving $41.00 per month is much better than losing over $300.00.

Another comparison to use is the life expectancy of the furnace and a vehicle. According to consumerreports.org, the average life of a car is eight years and 150,000 miles. IHS Markit estimates a life expectancy of 11.5 years. According to ASHRAE, the life expectancy of a furnace is 18 years.

Look how smart I am

Service technicians are sometimes known for their ego and being prima donnas. I urge you to keep your ego in check when dealing with customers. When I was first in sales, my boss used to make me take the service manager on sales calls. Huge mistake! He was one of the smartest technicians I ever met and was always eager to remind you of that. When I would bring him to the job site to evaluate the customer's system, he would find a way to insult the customer by saying something like, *"This system is junk and should be replaced."* Not a great way to start a sales call.

Best HVAC system is one no one notices

One of the frustrating things in our industry is a properly working heating and cooling system is almost anonymous. It is not like the shiny new truck parked outside. The only time it gets attention is when it is not operating correctly.

Energy reduction increases building value

One of the things you want to think about is the financial ramifications of what you are selling especially in commercial buildings. There is an industry rule of thumb which states every dollar cut from the bottom line of a building, increases the value of the building by a factor of ten. Consider this scenario:

You present two options for the customer. One is to install an exact replacement which will cost $30,000.00. The energy savings are essentially a wash as the new unit is identical to the existing unit. The other option is a more efficient system which will cost $35,000.00 and save the customer about $1,000 per year. The additional $5,000.00 investment would take five years to recover. Would it be wise for the customer to choose the more efficient system? Let us look at the numbers. The first option for the customer would be to take the $5,000.00 they are not paying and place it in the bank in an interest-bearing account paying 3% interest. *I realize this would never happen in real life but for comparison sake, let us pretend it does.* Assuming they leave the money there for twenty years, the $5,000 investment would become $9,030.56. The customer would get a nice 81% return on investment after 20-years and almost double their investment.

Now consider the option of purchasing the more efficient system. The customer will pay the additional $5,000.00, but the savings would be $1,000.00 per year. It would take roughly five years to pay back the extra investment. They would continue seeing the savings every year for the 20-year life of the equipment. If you consider the future value of the $1,000 savings every year for 20-years, it would equal $28,676.49. If we factor in the additional value of the building since we cut the costs by $1,000.00, the building would be worth an additional $10,000.00. Our $5,000.00 investment will generate $38,676.49 to the building owner after twenty years or a whopping 674% return on investment. This does not include any increases in fuel costs. Which option would you choose?

Read customer news

If you are specializing in a certain market niche, read the industry magazines or websites for that industry. It will help you understand what challenges your customer has and help you speak their language.

What is the cost?

Every once in a while, you will have a customer tell you to skip the sales presentation and say, *"Just tell me the price."* My sales trainer friend Ken Launikonis would tell the customer, *"If I tell you it will cost a million dollars but will generate ten million dollars, that would be a great deal. If I tell you it cost $10,000 and has absolutely no value, that would be a bad investment. I promise to get to the prices very quickly. Let me see whether my company is even a good fit for yours. Is that ok?"* Most clients will back down and let you continue.

How building comfort affects the building owner.

The comfort inside the building can affect the occupancy of the building. If the tenants are too cool or hot, they will leave. If you can increase the comfort inside the building, it affects the owner's bottom line. The cost of tenant turnover is very expensive for the building owner. The average cost of tenant turnover is said to be equal to three month's rent plus the lost rental income for the space.

Depreciation

When a customer purchases a piece of equipment for their building, the Internal Revenue Service will allow them to depreciate the value of the unit over a period of years. Residential rental properties can depreciate the unit's cost over 27 ½ years. Commercial buildings can depreciate the value of the equipment over 39 years. A basic way to understand depreciation is to take the cost of the unit and divide it by either the 27 ½ or 39 years. The amount can be deducted from the income for the building. Now I know what you are thinking, many of the new units will not last 39 years or even the 27 ½ years. Some building owners are leasing the equipment. In many instances, the entire lease payment is deductible. Take an accountant to lunch to get a better idea about this so you can help your customer.

It's how you say it

One of the words I like to avoid on any sales call is cost. I prefer using the word investment. Cost means money will leave their wallet and they will not get anything in return.

Investment means they will earn more money than they are paying you.

Ask for the order

When you are done with your sales call, you have to ask for the order. The customer is expecting it and many will not buy unless you ask for the order. Asking for the order may be something as simple as asking *"Are we ready to go ahead with this now?"* Some salespeople feel uncomfortable asking for the order directly. If that is the case, you could do indirectly. *"Our installation department is available on Monday or Wednesday. Which day works best for your schedule?"* If the client says either day, congratulations you got a sale. Another way of indirectly asking for the order is to ask: *"Would you like us to take the old furnace with us or leave it here?"* You know you will take the old furnace but it gauges whether they are ready to purchase.

Wow, great question

I like to say this when a customer asks me a thoughtful question. Now the question has to be pertinent but telling the customer they asked a great question makes them feel good about themselves. I will also say this when teaching a class. A different variation on this is to say, *"I'm glad you brought that up."* I would then answer the question.

How do you know when the customer is ready to buy?

The signs the customer is ready to buy is when they ask questions about the sales call. For example, they may say ask when could we schedule this? Or is the unit in stock?

ABC's

ABC or Always Be Closing was an acronym I learned early in my career and to me it seemed so adversarial. I never felt comfortable with the battle. Some sales consultants urge you to close at least seven times and that is way too much of a battle for me. I believe if you do a proper job of questioning, that is where you sell your product or service.

Types of sales closes

So where do we go from here? - I like using this as it throws it back to the customer. When done with my presentation, I will look at the customer and say, "So where do we go from here?" The customer in most instances will say "I guess we should get started."

Impending Event Close – This is one of my least favorite closes as it has been used and abused by many a sales person. Some sales people will use it as a way to pressure the customer to make a decision right away and the customers have seen and heard it all their lives. When R22 units were discontinued, some wholesale houses still had some left over. I was talking with a customer about a replacement condenser for his rental property, I used this close and told him I was not sure how long the units would be in stock. The customer gave me the order that day.

Ben Franklin Close – Ben Franklin would make a major decision by listing all the pros on one side of a sheet of paper and all the cons on the other side of sheet. He would look to see which side had more points and make the decision to go with the side with the more points. I will explain this process to the customer and suggest if Ben Franklin, one of the wisest people in history used it, perhaps it would work for them. I have used this close on analytical people and it has worked well. Start off by asking the customer to list any ideas opposed to going ahead with the sale. This catches the client off guard a bit and they usually smile and list the ideas holding them back. Once the negatives are listed, I will ask the client for reasons for going ahead. They may have thoughts I did not consider. Once done, I add my own reasons for going ahead and hopefully the positives will outnumber the negatives. After completing the lists, show it to the client and ask which one makes more sense. I hope you noticed the wording I used above. I had the customer list all the *ideas* opposed and then suggested the *reasons* for going ahead. A reason is much stronger than an idea. It is a subtle way of directing the sales call. A variation of this is to write the name of the sale on top of the page and draw a line down the center of the page from top to bottom. I hand the tablet to the customer and ask them to complete it and offer suggestions. In this way, the list is theirs and they are quicker to act.

Assumptive Close - This close assumes the customer will go ahead with the proposal. I make a mark on the proposal where the customer signs and push it to them and say *"I just need your authorization to get started."* And lean back in my seat and wait to see what happens. In most instances, the customer will sign the proposal.

Something free for today – Some people will offer a special deal to the customer if they sign today. I have had this backfire on me. I told a customer I could give them a humidifier if they decide to go ahead today. The customer decided not to go ahead with my proposal but called a few days later and said they reconsidered. They still wanted the humidifier. Now I was in a difficult position as I told them it was only a special that day so I would not have to make the trip back to them. If I agreed, It looks like I lied to the customer during the sales presentation. In this instance, I told them I had to check with the supplier and let them know. I told them I would call them right back. I waited about 5-10 minutes and called them back and said, *"I have good news. I can offer that to you."* It allows me to save face and looks like I am working for the customer's interest.

This or that – This is one of my favorite closes as I like offering a couple of options in my proposals, such as the good, better, or best options. I will ask the customer which option they are leaning toward.

Don't ask for signature

A little quirk of mine is never to ask a customer for their signature. It makes it sound so formal and legal and sometimes intimidates the customer. I like to ask them for their authorization or their autograph, which makes them feel special.

Let me check with...

I like using this sales tool as it makes it look like I am working on the customer's behalf with the *"evil"* factory, manufacturer, boss, or wholesaler. If the customer asks for better pricing or terms, I will tell them I will call on their behalf. I will also ask, *"If we can do this, will you be looking to go ahead?"*

In the middle of every difficulty lies opportunity. Albert Einstein

After the Sales Call

Did you actually save money?

Do you ever contact your client and follow-up after the sale? You should consider doing this as it will enforce to the customer they made a sound decision by choosing your firm. It will also let you know whether the customer really saved money by using your firm or product.

One of the things you could do is to review their energy usage and compare it to the year before the new system was installed. This can do two things, it will reassure the customer for using your firm and it could be a powerful sales tool for other customers. A simple way to estimate savings is to compare energy consumption and divide it by the yearly degree days. The degree days for your location are available on line through the weather bureau. Be sure to use the actual fuel consumption and not the cost of the fuel as it could fluctuate.

What if we did not have savings?

We sold boilers to a church and the business manager asked how much I thought they would save. I estimated a savings of 20%. After the first winter, I contacted the business manager and he said the church did not see any savings and his fuel costs actually rose. I had a hard time believing it as we replaced an old inefficient boiler with two new very efficient boilers. I asked for the utility bills for the last two years. During my research, I made two discoveries. The first was the price per cubic foot of gas rose substantially. The second cause was the degree days for the last winter were about 20% higher than the previous year. After I redid my calculations based on the actual conditions, the savings were closer to 30%. I explained it to my customer and he became an advocate for our company to the other churches in the area. If I had not done this, he would have believed our boilers cost him more to operate. Do you think he would have ever recommended us or our boilers?

Be careful guaranteeing savings

I would be hesitant guaranteeing savings for any customer. There are many variables which could affect your estimate. One of our customers had an antique pneumatic control system which did not work very well. What the school had instead was the Heat Miser, a custodian who would walk through the school and shut off the heat when he felt it was warm enough. Have you ever noticed how a sunny 50-degree day feels warm in winter but cold in summer? He would shut off the heat on those days. When the custodian retired, we installed a new control system. The customer asked me how much he thought we would save and I boasted we would save about 25%. The fuel costs rose and it was because my control system could not compete with the custodian walking through the building.

Evaluate the sales call

Right after the sales call, think about how it went and how it could have been better. I like to do this even if I sold it. I ask myself, what could I have done better? I would urge you to evaluate every call even if you sold it. It will make you a better sales person.

Always send a follow-up

One of the keys to sales is the number of contacts you have with the clients. It is said people only remember you after seven contacts. After the sales presentation, I may send a sheet to the customer of something we spoke about in the sales process. For example, if the client golfs, I may print a sheet about the local golf courses and personalize it by writing something on the sheet and mailing it to the person with my card.

The power of Thank You Cards

I really enjoy using handwritten thank you notes in my business. I believe it separates me from my competition. Think about your own feelings when you open your mail box and see a handwritten envelope. It sure beats the feeling of receiving a bill. I will send out a thank you note when I meet with a client the first time or if someone does something nice for me. I like using thank you cards with our logo and company name on the cards so the customer remembers us.

Make notes

I like to stop for a coffee right after the sales call and make notes about the customer and the sales call while it is fresh in my mind. I like to note personal information about the customer in the file. I suggest you go to a place other than the office to review the sales call as once you get back to the office there are usually many distractions and you will not have a chance to write your notes.

Typical Sales Process

The following page is a diagram showing the typical sales process. It can be used as a guide for building your own sales presentation.

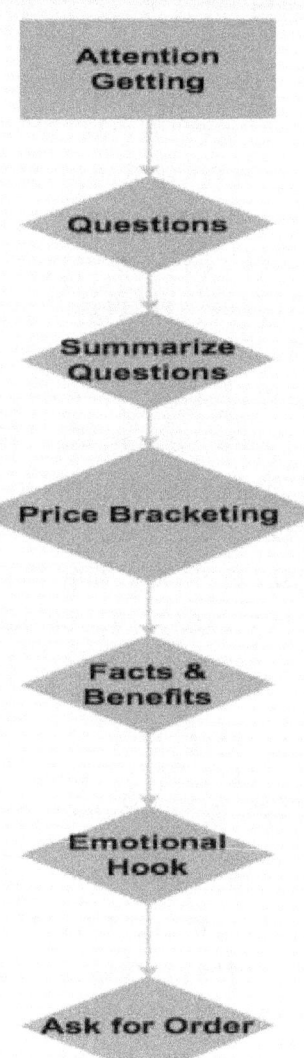

Keeping your customers

You are my favorite

My friend has four daughters and he would play a little trick on the daughters. He would have a Daddy Daughter date where he took each one out separately for ice cream or lunch. During those times, he would tell the daughter, "*You are my favorite. Don't tell your sisters.*" This was his way of making each daughter feel special regardless of the birth order. The reason I am sharing this story with you is this would be a great idea for your customers. Make each customer feel they are the most important customer, much more important than any other client. If you do that, the customer will appreciate the service and be hesitant to leave. In case you were wondering, the secret got out with my friend and each daughter found out what he did. They still loved him.

Should you give gifts to clients or prospects?

This is a slippery slope as sometimes the line is blurred between what is a gift and what is a payout. We give engraved pens or coffee mugs to customers and sometimes take a customer to lunch. I have been asked for gifts and or bribes and chose not to do it. A local school district representative asked me to install a new furnace, AC unit, thermostat, and humidifier in his house and we refused.

The employee was fired after it was discovered he did this to several companies. Many facilities limit the cost of gifts to $25.00 or less.

Know your customers

To keep your customers; you have to know them. While it is assumed you need to provide the level of customer service they require, you have to know what they like and expect as far as service. Some like to know everything you do while others trust you to do the job properly. I think to keep them as customers, you have to know them as a person. For example, one contractor I know sends out birthday cards to every customer.

One of the things I like to do is to scan the newspapers daily to see if any of my customers are in the news. If so, I will send a link to the story with a short note. For example, one of my clients was quoted in an article. I sent him a link and said: "*I enjoyed the article.*" He could not believe anyone saw it and thanked me for the feedback.

One of my niche markets is colleges and universities. I like to scan the local and national news about my markets. If I find something interesting, I will forward a link to my clients with a short note. It is all about contact occurrences.

Keeping your existing clients

The cost to get a new customer is between four and ten times the cost to keep an existing client. We have some turnover in our industry and I urge you to do your best to keep the existing clients. If you have shopped at any large company, I will wager you have been asked by almost every cashier to go on line and fill in their questionnaire. These organizations must believe it is worthwhile because they all do it. Every company needs to see how they are doing in the customer's mind. Some HVAC companies have survey cards on self-addressed stamped post cards and the technicians leave them on the job site.

The following are the leading causes of why customers leave:

68% leave because of poor attitude or indifference on the part of the service provider.

14% leave because of product dissatisfaction

9% leave for competitive reasons

5 % leave because of other friends

3% move away

1% die

Remember me

Try not to take it personally but people tend to forget their heating and cooling company. If they are lucky, we only see them a few times a year. In many instances, we only see them every few years. I like to use stickers and place one on each piece of equipment. In this way, they can see your company name and number. I like larger stickers as they can cover the sticker from the previous service company. A friend pays a spiff for any competitors stickers the techs brought back.

Many thermostats allow you to program your company name and telephone number so it flashes when the unit requires service. If you do enough replacements, you can purchase thermostats with your logo and telephone number screen printed on it.

Who is next in line?

While you may have a good relationship with your building contact, you should get to know the others in the facility in case your contact ever leaves. We used to service the buildings for a property manager of a local real estate management firm. When she left, her assistant took over. I had always been respectful of the assistant. We kept the business and the assistant eventually was responsible for more buildings. As a result, our business increased. Conversely, we had a good relationship with a property manager and a not so good relationship with his assistant. When our contact left, our business ended abruptly. He brought in his people.

If your contact leaves, they will most likely go to another organization where they could use your service. Keep in touch with your contacts. One of my customers was let go from her job and I sent her a nice note saying how much I enjoyed working with her. I also forwarded any job opportunities I saw posted. She started a new job at a larger organization and remembered us and brought us in for service.

Donuts can clinch the deal

I like to bring donuts sometimes to my commercial customers as a way of saying thanks for the business. I will attach one of my business cards to the box and drop it off. You will be amazed at how some people remember it.

You must be the change you wish to see in the world. Mahatma Gandhi

Industry Stories

Some customers enjoy stories and they can help you close more sales. The following are some industry stories I use on sales calls:

The Carbon Club

In the 1800's, steam boilers were being installed in homes at a record pace. Boiler accidents and explosions were occurring at a record pace as well. It was like the wild west back then as there were no rules or codes about steam systems. Some installers would install high-pressure steam systems and it was common to have residential systems operating on 60-80 psig. Boilers did not have the safety controls they have today and there were explosions almost daily. In December 1899, a group of boiler manufacturers met and decided on a couple of standards we still see in use today. The first standard was low-pressure steam systems were designed to operate on 2 Psig steam pressure. If you see a comfort low-pressure steam system operating at pressures higher than that, there could be system issues. The second standard they adopted was using 180 degrees as the design temperature for hydronic systems.

The Spanish Flu

A global pandemic hit the world in 1918 and killed between 20-40 million people worldwide. People showing symptoms in the morning were dead by evening in many cases. The puzzling part was no one knew the cause or remedy for the flu. It simply ran its course. Many believed it was caused due to insufficient air to the home. It was recommended to sleep with your windows wide open. Heating systems were designed to heat the home with the windows wide open on the coldest day of the year with the wind blowing.

How the Living Room got its name

During the height of the Spanish Flu, dead family members were laid out in the front room of the home. It was known as the Death Room. Once the Spanish Flu subsided, the magazine Ladies Home Journal suggested we call the room the Living Room and the name stuck.

Sizing a steam system.

While furnace and air conditioners are sized according to the heat loss or gain of the home or building, low-pressure steam systems are sized by the connected load. To properly size a replacement steam boiler, the installer or designer should add all the radiators and piping and size the boiler for that. How does the building owner know if the boiler is large enough to heat the building in the middle of winter? They used to have a test to for it. The installer would have to heat the house very warm and then document it. For example, if the boiler were started on a day when the outside air temperature was 80 degrees F, the installer would have to heat the house up to 125.6 degrees F. If the outside air temperature was 50 degrees on the day of the startup, they would have to heat the house up to 104.8 degrees F. Now they have lawyers to make sure the system is sized correctly.

Did you know?

Natural gas is normally odorless and colorless. Mercaptan is added to the natural gas piping to allow it to be detected by odor. It is an organic substance comprised of carbon, hydrogen, sulfur. It occurs in humans and is one of the chemicals responsible for bad breath or flatulence.

Let us not pray to be sheltered from dangers but to be fearless when facing them. Rabindranath Tagore

Professional Organizations to find customers

Association of School Business Officials www.asbointl.org

APPA Association of Physical Plant Administrators www.appa.org

BOMA Building Owners and Managers Association www.boma.org

International Facility Managers Association www.ifma.org

National Association of Power Engineers www.powerengineers.com

National Real Estate Investors Association www.nationalreia.org

ABOUT THE AUTHOR

Ray Wohlfarth is president of Fire & Ice Heating & Cooling in Pittsburgh, PA. He is the author of five other books. Four books are about boilers and the fifth is a children's book. The books are as follows:

Lessons Learned in a Boiler Room
Lessons Learned: Connecting New Boilers to Old Pipes
Lessons Learned Servicing Boilers
Lessons Learned Brewing with Steam
Farmer Franks Forgetful Flock
He is also a columnist for several industry trade magazines. Ray can be reached at Ray@fireiceheat.com